Environmental Footprints and Eco-design of Products and Processes

Series Editor

Subramanian Senthilkannan Muthu, Head of Sustainability - SgT Group and API, Hong Kong, Kowloon, Hong Kong

Indexed by Scopus

This series aims to broadly cover all the aspects related to environmental assessment of products, development of environmental and ecological indicators and eco-design of various products and processes. Below are the areas fall under the aims and scope of this series, but not limited to: Environmental Life Cycle Assessment; Social Life Cycle Assessment; Organizational and Product Carbon Footprints; Ecological, Energy and Water Footprints; Life cycle costing; Environmental and sustainable indicators; Environmental impact assessment methods and tools; Eco-design (sustainable design) aspects and tools; Biodegradation studies; Recycling; Solid waste management; Environmental and social audits; Green Purchasing and tools; Product environmental footprints; Environmental management standards and regulations; Eco-labels; Green Claims and green washing; Assessment of sustainability aspects.

More information about this series at http://www.springer.com/series/13340

Subramanian Senthilkannan Muthu
Editor

The Water–Energy–Food Nexus

Concept and Assessments

 Springer

Editor
Subramanian Senthilkannan Muthu
Head of Sustainability
SgT Group and API
Kowloon, Hong Kong

ISSN 2345-7651   ISSN 2345-766X   (electronic)
Environmental Footprints and Eco-design of Products and Processes
ISBN 978-981-16-0241-2   ISBN 978-981-16-0239-9   (eBook)
https://doi.org/10.1007/978-981-16-0239-9

This Springer imprint is published by the registered company Springer Nature Singapore Pte Ltd.
The registered company address is: 152 Beach Road, #21-01/04 Gateway East, Singapore 189721,
Singapore

This book is dedicated to:

The lotus feet of my beloved Lord Pazhaniandavar

My beloved late Father

My beloved Mother

My beloved Wife Karpagam and Daughters—Anu and Karthika

My beloved Brother—Raghavan

Everyone working with various industrial sectors to make our planet earth SUSTAINABLE

Contents

About the Editor

Dr. Subramanian Senthilkannan Muthu currently works for SgT Group as Head of Sustainability and is based out of Hong Kong. He earned his Ph.D. from The Hong Kong Polytechnic University and is a renowned expert in the areas of environmental sustainability in textiles and clothing supply chain, product life cycle assessment (LCA) and product carbon footprint assessment (PCF) in various industrial sectors. He has five years of industrial experience in textile manufacturing, research and development and textile testing and over a decade of experience in life cycle assessment (LCA), carbon and ecological footprint assessment of various consumer products. He has published more than 100 research publications, written numerous chapters and authored/edited over 100 books in the areas of carbon footprint, recycling, environmental assessment and environmental sustainability.

Reflections About the Food–Energy–Water Nexus in a World Without Economic Growth—A Dynamic Multinational CGE Model-Based Thought Experiment

Holger Schlör⬛, Stefanie Schubert⬛, and Sandra Venghaus⬛

Abstract In response to the intensifying need to mitigate climate change and reduce the pressures on the earth's natural resources, alternative economic (post growth) concepts begin to outstrip their role as niche concepts and are now frequently hypothesized to provide inevitable contributions to solving today's sustainability challenges. Almost half a decade ago, Meadows et al. (The limits to growth. Universe Books, New York, 1972 [51]) instigated the discussion about "the limits to growth", an idea which was later supported by the Brundtland Commission's (World Commission on Environment and Development (WCED) in Our common future. Oxford University Press, Oxford, New York, 1987 [94]) call for new growth models. More recently, authors such as Stiglitz et al. (Report by the commission on the measurement of economic performance and social progress. Paris, 2009 [74]; Mismeasuring our lives: why GDP doesn't add up. New Press, New York, 2010 [73]) or Tim Jackson have demanded a complete redefinition of prosperity with the objective to decouple human welfare from the material impact on the environment (Jackson in Prosperity without growth. Earthscan, London, 2009 [37]). This discussion about new models of prosperity often finds itself under the label of the "green economy" or the "food–energy–water nexus" (FEW nexus) addressing the core sectors of this transformation process. In the developed model, these new ideas are integrated in an intertemporal dynamic multinational general equilibrium model (GE model). The GE model consists of four countries (A, B, C, D) with three economic sectors (FEW, households, industry) each. We discuss the economic effects of our GE model approach in four growth

H. Schlör (✉) · S. Venghaus
IEK-STE, Forschungszentrum Jülich, Jülich, Germany
e-mail: h.schloer@fz-juelich.de

S. Venghaus
e-mail: s.venghaus@fz-juelich.de

S. Schubert
SRH, Heidelberg, Germany
e-mail: Stefanie.Schubert@srh.de

S. Venghaus
School of Business and Economics, RWTH Aachen University, Aachen, Germany

© The Author(s), under exclusive license to Springer Nature Singapore Pte Ltd. 2021
S. S. Muthu (ed.), *The Water–Energy–Food Nexus*,
Environmental Footprints and Eco-design of Products and Processes,
https://doi.org/10.1007/978-981-16-0239-9_1

scenarios for the four countries: Country *A* follows a zero-growth scenario, countries *B* and *C* grow by moderate rates of 1.2% and 1.9%, respectively, and country *D* is on a de-growth pathway (−1.3%). This model approach reveals the possible socio-economic consequences and alterations of various growth models for the FEW nexus sector, as well as the other economic sectors of the four countries.

Keywords Food–energy–water nexus · General equilibrium model · History of economic growth · Growth models

1 Introduction

As it becomes undeniably clear that the window of opportunity to stop the unrestricted and accelerating climate change is about to close [22, 30, 40, 95], Porter proposed a thought experiment to leave the current global economic model behind [59]. He recommended to discuss and imagine instead a world without economic growth [59] to reduce CO_2-emissions as demanded by the IPCC [36] and the World Meteorological Organization (WMO) [87] to limit global warming to 1.5 °C [36].

The current scientific and political discussions increasingly support this idea and challenges the necessity and plausibility of the prevailing economic growth model [37, 38, 50, 77, 89], raising instead questions as the following:

- Is "the burning of fossil fuels … the price of continued economic growth [88]", or could a world without economic growth lead to CO_2-reductions [88]?
- Is economic growth a natural phenomenon [72]?
- And if so, are economic growth and sustainability mutually exclusive, or is it possible to find a balance between economic growth and the requirements of sustainability?

Taking a look at the concept's historical emergence, it becomes apparent that the discussion about alternative economic growth models started nearly 60 years ago with the concern about a 'silent spring' [16], referring to the general conditions of the global environment. The book leads to the 1972 United Nations Conference on the Human Environment in Stockholm and the Club of Rome report in the same year. This discussion was mainly influenced by Daly's "Steady State Economics" [18], Georgescu-Roegen's fundamental paper "The Entropy Law and the Economic Process" [29] about the biophysical limits of growth, Boulding's paper on "The Economics of the Coming Spaceship Earth [12]". Hardin's essay "Tragedy of the Commons" started the discussion about institutional requests for a sustainable development [31], and Schumacher discussed a new institutional economic framework published in his book "Small is Beautiful" [68]. These books built the basis for Ostrom's book about "Governing the Commons" [56] who added new institutional ideas on sustainable development.

Those concepts can be considered as the economic foundation of sustainability science and have influenced the UNEP ideas of a green economy [81] and of a circular economy [91], providing a cornerstone of new economic growth models [20, 91]. In 1983, the Brundtland Commission was established, and proposed in its report that "environmental challenges arise both from the lack of development and from the unintended consequences of some forms of economic growth" [94]. And at the Bonn2011 Conference "The Water, Energy and Food Security Nexus", the UN and further international organizations discuss the nexus approach "that integrates management and governance across sectors and scales [34]." It is argued that the FEW nexus concept is a central element in the transition towards a green economy representing thereby the key sectors of sustainable development [27].

Based on this preliminary work, the UN in 2015 further asserted in their resolution A/RES/70/1 "Transforming our World: The 2020 Agenda for Sustainable Development [83]" that the world community must create conditions for sustainable, inclusive and sustained economic growth [83]. Whereas the UN did not see a fundamental contradiction between economic growth and sustainability, and thus adopted Goal 8 (Promote sustained, inclusive and sustainable economic growth, full and productive employment and decent work for all) as one of its Sustainable Development Goals (SDGs) [83], the organizers and participants of the Post-Growth Conference in Brussels instead propose in the Post-Growth Open Letter to EU institutions the end of the growth dependency of Europe [60].

Addressing this controversial debate, the objective of our model is to contribute to the discussion about the characteristics of a post-growth economy. The remainder of this paper is structured as follows: First, we will briefly reflect on the history of economic growth, addressing also the economic growth effect of the current Corona pandemic (Sect. 2). In Sect. 3, the economic foundation of the neoclassical growth models will be presented. In Sect. 4, the FEW nexus economic sectors will be explained as well as its relationship to the main economic indicators. In Sect. 5, the different economic growth scenarios inspired by the works of Jackson [37, 38], the Club of Rome [50, 62], Victor [85, 86], Weitzman [89] and Trainer [77] will be discussed. In Sect. 6, the consequences of our results will be discussed.

2 Economic Growth in History

The analysis is started based on Porter's considerations [59] with a short summary of the history of economic growth and recent economic developments caused by the Corona pandemic.

2.1 Historical Deliberations

Reverting to the question as to whether the striving for low or zero economic growth is actually something new, Porter claims it is "hard to imagine now, but humanity made do with little or no economic growth for thousands of years" [59], as the data of Maddison verified [44–48]. The Economist wrote about the pioneering work of Maddison: "He believed that the 'pace and pattern' of economic activity had deep historical roots. Economies, he thought, do not 'take off', as if from nowhere [76]." And the New York Times characterizes Maddison as a scientist, who "devoted his life to forecasting the past [61]." In the following, it will be analysed how the current economic growth emerged, presenting some highlights of Maddison's forecast of the past and its historical roots.

Comparing data across such a long period requires the consideration of relevant concurrent developments. In this context, especially the development of the world population is of interest. As Fig. 1 shows, there was no significant change in the world population during the first thousand years and only a very moderate growth of population until 1700.

From 1700 onwards, however, population growth had significantly accelerated for roughly 200 years, before it finally took off in the twentieth century.

The data of Table 1 further shows that until 1820 GDP and the world population developed relatively similarly. However, between 1820 and 1870 the gap became larger. By 1870, the population had risen roughly by a factor of six as compared to year 1, whereas GDP had increased by more than a factor of 10. This split further increased over the next 140 years. Considering the full period from year 1 until 2008,

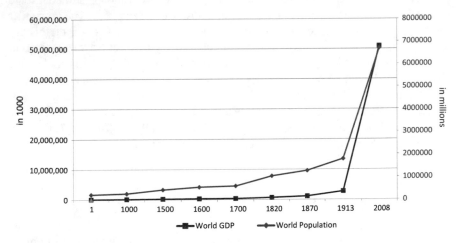

Fig. 1 History of world population and GDP—year 1 until 2008. *Source* Own calculation (2020) based on Maddison (Groningen Growth and Development Centre (GGDC)) [48]. *GDP: Million 1990 International Geary-Khamis dollars

Table 1 Increase of world population and world GDP—year 1 = 100

	1820/1	1870/1	1913/1	2008/1
Population	461.3	564.9	794.0	2964.7
GDP	658.0	1052.8	2593.1	48,361.4

Source Own calculation (2020), based on Maddison [48]

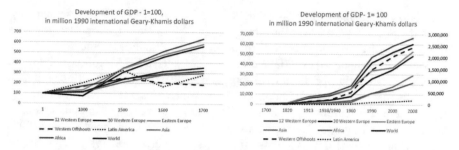

Fig. 2 Development of GDP *Source* Own calculation (2020), based on Maddison [48]

Table 2 Per capita GDP—year 1 = 100 (1990 international Geary-Khamis dollars)

Region\Year	1	1000
Total 12 Western Europe	100	70.87
Italy	100	55.60
Total 30 Western Europe	100	74.18
Total 7 East European countries	100	97.14
Mexico	100	100.00
Total Latin America	100	100.00
Total Asia	100	103.14
Total Africa	100	89.93
Egypt	100	83.33
World average	100	97.14

Source Own calculation (2020) based on Maddison [45]

the world population grew by a factor of roughly 30, whereas the average global GDP as measured in Geary–Khamis Dollar over the same period grew by a factor of 484.

Figure 2a, b show how this World GDP is distributed between the different regions of the world. Figure 2a reveals that the GDP did not change significantly until 1700 in comparison to the years thereafter.

The GDP in the 12 West European states (Austria, Belgium, Denmark, Finland, France, Germany, Italy, Netherlands, Norway, Sweden, Switzerland, UK) increased roughly sixfold over the 1700 years as Table 2 shows: The highest increase in all analysed regions, followed by Eastern Europe (Albania, Bulgaria, Czechoslovakia, Hungary, Poland, Romania, Yugoslavia) and by all 30 Western Europe[1] states.

The lowest increase occurred in the Western Offshoots (Australia, Canada, New Zealand, USA). In Latin America, Asia and Africa, the economy grew nearly at the same rate until 1700. So, overall, during this period the world economy grew by a factor of 3.5. This picture, however, changed dramatically after 1700.

---

[1]The 12 states of Western Europe plus Ireland, Greece, Portugal, Spain, and plus 14 small west European countries.

Between 1820 and 1938/1940 (Fig. 2b), the GDP increased rather continuously. After the Second World War, however, a significant break occurred leading to a massive increase in economic performance. The GDP increased most strongly in the Western Offshoots (Australia, New Zealand, Canada, USA), followed by Latin America, Western Europe, Eastern Europe, and Africa and Asia. The strong increase especially for the Western Offshoots and Latin America is rooted mainly in the fact that in these two regions the start level in year 1 was very low in comparison to Asia, Africa and Europe. The figure also shows that economic growth after 1700 is a product of the industrial revolution—initiated by the invention of the steam engine by James Watt in 1769 and the subsequent rise of fossil energy consumption [70, 71].

Whereas the figures depict the developments of the world populations and of the GDP for the various world regions, in order to derive a clear picture of the welfare development over the last 2000 years, we further need to analyse the per capita GDP growth: the growth of the individual welfare based on the economic development, as Table 2 shows.

The historical data of Maddison (Table 2) reveals that the world average per capita GDP in the first millennium decreased slightly from the dawn of the Christian era to the end of the first millennium. During this period, especially Italy experienced a dramatic decrease of 45% of its per capita GDP. After the collapse of the Roman Empire, Maddison detected "a significant decline in per capita income in all the West European provinces, with the biggest drop in Italy where population fell by a third and per capita income by nearly half [47]." This general trend was caused by the vanished Roman Empire and its institutions, and holds for all Western European countries, although slightly less pronounced than in Italy. Similar developments were found for Egypt, whereas the development for Africa was less drastic. A slight increase in per capita GDP during this period was only found for Asia [48]. In Mexico and Latin America, the per capita GDP remained largely unchanged for most of those one thousand years, and for the group of the 30 Western European economies the economic growth (per capita GDP) declined by about 26% until 1000.

Following a series of technological innovations (light bulb, transistor, telegraph, car, aircraft) [47], the per capita GDP more than doubled until 1820 and tripled from 1820 until 1913 at the eve of World War I (Fig. 3). The decline of the per capita GDP in Eastern Europe was very moderate until 1000. After the year 1000, the per capita GDP increased there, also, although strongly than in Western Europe.

The Western Offshoots (Australia, Canada, New Zealand, and USA) began to grow a little later around 1700 [48]. However, their growth rates rose much more rapidly, outperforming Western Europe, tripling already in 1820. By 1913, the region had increased its per capita GDP by a factor of 13 compared to the year 1000. The GDP per capita in Africa decreased by roughly 11% between 1 and 1820, and then increased slowly until 1913. In Asia, the per capita GDP increased slowly by roughly 50% between 1 and 1913, and economic growth in Latin America and Eastern Europe began after 1870. However, these regions grew slower than Western Europe and the Western Offshoots. Economic growth started in the Western Offshoots and in Western Europe after 1700 due to the starting industrialization and accelerated in the coming centuries and spread over to the other world regions [44–47].

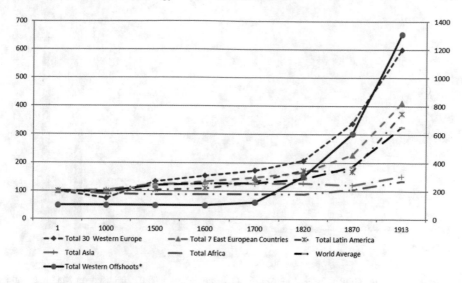

Fig. 3 Growth of GDP/capita − 1 = 100, *Australia, Canada, New Zealand, USA. *Source* Own calculations (2020), based on Maddison (Groningen Growth and Development Centre (GGDC)) [48], in 1990 International Geary-Khamis dollars

This historic overview demonstrates well that for nearly 1700 years, low or no per capita GDP growth was a normal global development, as Table 3 shows.

Between year 1 and 1913, the economic output (GDP/capita) of Western Europe increased by just about 0.09% annually. In Africa and Asia, GDP/per capita increased by roughly 0.02% per year, whereas Europe and Latin America grew by similar rates (+0.07%). Only the Western Offshoots (USA, New Zealand, Australia, Canada) grew at slightly higher rates (+0.13%). Accordingly, across all regions the per capita GDP grew on average by about 0.06% annually until 1913, as Table 3 shows.

In the following, we will focus on three countries (China, Germany, USA) in Asia, Europe and the Western Offshoots in further detail in order to elaborate the welfare effects exemplarily.

Table 3 Average annual GDP/capita growth rate between

	1 and 1820 (%)	1 and 1913 (%)
30 Western Europe states	0.04	0.09
7 East European countries	0.03	0.07
Latin America	0.03	0.07
Western offshoots	0.06	0.13
Asia	0.01	0.02
Africa	−0.01	0.02
World average	0.02	0.06

Source Own calculations (2020) based on Maddison [48]

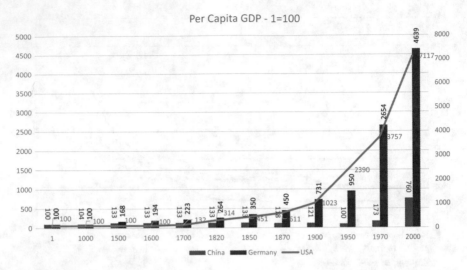

Fig. 4 Per capita GDP − 1 = 100. *Source* Own calculations (2020), based on Maddison and Groningen Growth and Development Centre (GGDC) [48]

In China (Fig. 4), per capita GDP remained more or less the same until 1000 and then rose slowly until 1500, before it stagnated until 1850, and then shrunk again until 1950 to the level of year 1. During this middle period, there was no economic growth and per capita GDP remained constant for nearly 500 years. In Germany, instead, per capita GDP remained roughly constant for the first 1000 years, before it more than doubled until 1700. Then the per capita GDP increased constantly until 2000. In the USA, there was almost no per capita GDP growth for nearly 1700 years. It started to grow after 1820 and then the economic performance rose sharply with per capita GDP increasing constantly until 2000.

Given the examples of China (with almost no economic growth per capita for 2000 years) and the USA (with also almost no per capita welfare growth for nearly 1700 years), we can safely conclude that economic growth is not a natural constant. It, instead, depends on the socio-economic circumstances. However, de-growth is not only a historical event in Italy and China, as the economic effects of the Corona pandemic currently show.

2.2 Corona Virus—A Historical Economic Growth Break-In

The COVID-19 pandemic and the related governmental containment measures caused a collapse of the world trade and many other economic activities (tourism, capital flows, remittance) [2]. The global shutdowns initiated by the governments around the world caused a tremendous downturn of the global economy, as Table 4 based on the estimates by the World Bank shows [93].

Table 4 Coronavirus impact on economic growth, 2020

	2017	2018	2019	2020e	Difference 2019/2020
World	3.3	3.0	2.4	−5.2	−7.6
Advanced economies	2.5	2.1	1.6	−7.0	−8.6
Emerging markets	4.5	4.3	3.5	−2.5	−6.0
High-income countries	2.4	2.2	1.7	−6.8	−8.5
Developing countries	4.8	4.4	3.7	−2.4	−6.1
Low-income countries	5.4	5.8	5.0	1.0	−4.0
BRICS	5.3	5.3	4.7	−1.7	−6.4
World trade volume	5.9	4.0	0.8	−13.4	−14.2

Source World Bank [93], e = estimations of World Bank

The global economic growth for 2020 is estimated by the World Bank to shrink by nearly 8%, with especially high values for the advanced economies (−8.6%) and high-income countries (−8.5%). These countries are facing severe reductions in their economic activities. Furthermore, also the emerging markets, developing countries and low-income countries will be faced with reductions in their economic activities by around −6 and −4%. Especially dramatic is the decline of the world trade volume by about −14.2%.

The U.S. Bureau of Economic Analysis (BEA) also analysed the effect of COVID-19 on the US economy. The BEA stated that the "Real gross domestic product (GDP) decreased at an annual rate of 32.9% in the second quarter of 2020 [78]" (Fig. 4), according to the "advance" estimate released by the Bureau of Economic Analysis. In the first quarter, real GDP fell by 5.0% (Fig. 5).

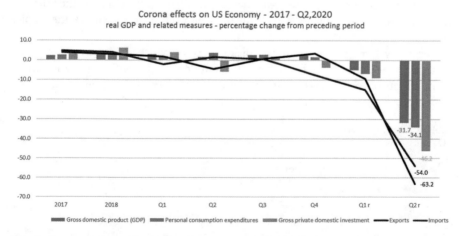

Fig. 5 Corona effect on US economy. *Source* Own graphic (2020) based on https://www.bea.gov/news/2020/gross-domestic-product-2nd-quarter-2020-second-estimate-corporate-profits-2nd-quarter

The Bureau explained "the decline in second quarter GDP reflected the response to COVID-19, as "stay-at-home" orders issued in March and April were partially lifted in some areas of the country in May and June, and government pandemic assistance payments were distributed to households and businesses [78]." The BEA added that the bureau currently does not fully reveal the total economic effects of the COVID-19 pandemic on the US economy [78].

The decline of the real GDP reflects the break-in of the personal consumption expenditures of the US consumers by about -34.6% in the second quarter of 2020, whereas the gross private domestic investments crashed by -49%. This collapse also has an effect on the exports and imports of the US economy. The US imports are breaking down by -56.7%, the exports even by -64.1%. This decline of the US economy is part of the larger, worldwide economic decline [55]. This global development also has an effect on the development of the global CO_2-emissions. The severe governmental reactions of countries around the world in response to the COVID-19 pandemic and the essential global shutdown have led to a dramatic reduction in energy demand around the world [42], especially caused by the industrial closures and changed consumption patterns (Table 5).

Le Quéré and her colleagues detected that CO_2-emissions had decreased globally by -17% by early April in comparison to the mean 2019 levels. In China, the CO_2-emissions were reduced by 242 $MtCO_2$ due to the lockdown measures, and in the USA by around 207 $MtCO_2$. Both countries are responsible for a global CO_2-reduction of 43%, whereas an additional CO_2-reduction of about 21% has been calculated for the European Union and India. In total, the four major economies account for roughly 64% of the global CO_2-emission reductions caused by the governmental anti-pandemic measures. UN General Secretary Gutteres and the European Environment Agency argue that the economic effects of the corona shutdown caused a breather of the unrestricted and accelerating climate change, opening a window opportunity to stop climate change before it is totally closed [22, 30]. However, at the same time this breather seems to be counteracted by the tremendous fires in California, the Arctic, the Amazon, and the Pantanal [4, 52, 92].

The foray through the economic history has shown that zero-growth or even negative growth rates are not unnatural. In the following, we will pursue an analysis of the effects of when zero or negative growth is planned with full intent as part of

Table 5 Change in fossil CO_2 emissions (1.1.2020–30.4.2020)

	In $MtCO_2$	In % of global
China	−242	23
US	−207	20
EU28	−123	12
India	−98	9
ROW	−378	36
Global	−1048	100

Source Le Quéré et al. [42] and own calculations (2020), ROW = Rest-of-the-World

an economies strategy, for example in order to reduce CO_2-emissions, and how this idea translates into all the different layers of the economy especially in the FEW nexus sectors.

To answer this question, we developed an economic model. In the following, we will lay the foundation of the neoclassical model of economic growth.

3 Modelling Economic Growth and the FEW Nexus: Theoretical Foundation

We will discuss the most important neoclassical growth model: the Solow Model [72]. Solow's article inspired researchers conducting economic investigations on economic growth, which have a significant impact on economic policy around the world [41].

3.1 Solow Model[2]

The Solow Model was developed by Robert Solow and at the same time by Trevor Swan in 1956 [1, 72, 75]. It is therefore also called the Solow–Swan model [41]. The Solow model is the starting point of the neoclassical growth theory and a further development of the Keynesian economic growth model of the Harrod–Domar model [19, 32, 72].

Solow and Swan assume that a fraction of the production output is not used for consumption of the households but for investments. The annual investments in the modernization of the economy increase the capital stock of the Solow economy to even out the depreciation of the capital stock [41]. "The capital will accumulate as its marginal productivity is higher than the rate of depreciation. But with increasing levels of the capital stock, the difference melts away until the marginal productivity of capital is equal to the rate of capital depreciation [41]." Thus, the capital accumulation ends.

Labour is the second production factor of the Solow model. The technological state of the country affects also the level of production and its long-term economic growth perspective. Solow defines his model economy as an economic unit of all production and consumption activities. In his basic model, Solow assumes that there is no state and that all prices are constant [72]. Based on these assumptions made by Solow, we will present the Solow growth model with and without technological change.

---

[2]This chapter is based on the detailed description of the Solow model by Lange and Acemoglu [1, 41]. Lange also presented a detailed analysis of the Keynesian and Marxian growth models [41].

For the Solow model economy, we first assume that there is no technological change in the economy. Against this background, the following production function emerges [41], representing the technological possibilities of Solow's model economy [72] based on capital (K) and labour (L), (α) and $(1 - \alpha)$ represent the output elasticity of the production function.

$$F(K_t, L_t) = K_t^\alpha \cdot L_t^{1-\alpha} \tag{1}$$

The result is a new production function for the Solow economy:

$$F(K_{t+1}, L_{t+1}) = (I_t - \delta K_t)^\alpha \cdot (1 + g_L)^{1-\alpha} \tag{2}$$

I_t = net investments,

which in turn results in:

$$F(K_{t+1}, L_{t+1}) = (K_t + sY_t - \delta K_t)^\alpha \cdot (1 + g_L) \cdot L_t^{(1-\alpha)}. \tag{3}$$

Thus, in this Solow model approach, growth depends on the savings rate (s), depreciation rate (δ), change in labour input (g_L), the level of input of production factors, the output elasticities (α) and the existing unchanged technology.

If one now extends the Solow model and introduces technological progress (T) into the model, then the technical progress is inserted into the existing production function, resulting in the following production function [41]:

$$F(K_t, L_t, T_T) = T_t \cdot K^\alpha \cdot L^{(1-\alpha)} \tag{4}$$

For the Solow model economy, it is now assumed that technological progress is growing exogenously g_T, and thus results in Eq. (5) [41]:

$$F(K_{t+1}, L_{t+1}, T_{t+1}) = (1 + g_T)T_t \cdot (K_t + sY_t - \delta K_t)^\alpha \cdot (1 + g_L) \cdot L_t^{(1-\alpha)} \tag{5}$$

Thus, the growth of the Solow-designed economy is based on the available technology of the country and the technological progress the country can enable. So that we can summarize, based on Solow and the Lange interpretation, that capital accumulates in the country, and a continuous increase of the technological efficiency results finally in the economic growth of the country. On the other side, a zero-growth scenario is then based on a constant capital stock respectively declining capital stock based on the annual depreciation of the capital stock, a constant labour supply and unchanged technology [41].

The capital stock remains constant if the economic saving corresponds to the depreciation of the capital stock and thus the net capital growth is zero. Labour supply is in turn constant when the change in hours worked corresponds to the change in the labour supply of the population and there is no technological progress. Now, if one of these factors develops positively, the other two factors, or at least one of them, would have to develop negatively to compensate for the positive development [41].

Lange has shown [41] that the Solow growth model based on the neoclassical theory can produce a stable zero-growth scenario of the Solow-styled economy.

He defines the necessary assumptions for this scenario [41]:

1. an unchanged technological level of the economy,
2. no unemployment because wages are completely flexible and if the demand for labour declines, wages can also fall to allow the demand for labour to rise again but this assumption has severe social implications,
3. the capital stock of the Solow economy is unaltered. The investments of the economy are only necessary to compensate for depreciation of the existing capital. The savings rate of the model economy should correspond with the depreciation rate and the necessary investments to even the depreciation of the capital stock [41].

3.2 The Relation of Interest Rate and Time Preference

The previous discussion has shown that savings are important for economic growth. Hence, we will discuss the nature of the interest rate and its relation to the time preference [14, 15, 25, 26].

Starting point of this analysis are the economic decisions of the households. The household has to decide based on its preference order which part of its income (I) the household uses for consumption (C) today and which part of its income is saved (S) for further consumption. The saved income can be used by the household for future consumption so that an intertemporal benefit can be achieved [90]. If households renounce current consumption (C_0), the interest on savings (S_0) increases the possible future consumption (C_1). The interest rate (r) is the price for the current consumption renunciation and the postponement of consumption into the future [67, 90].

These assumptions result in the following income equation[3]:

$$I_0 = C_0 + S_0 \qquad (6)$$

For the next period, the interest on savings is available to the household as additional consumption (C_1):

$$S_0(1 + r) = C_1 \qquad (7)$$

If one solves the above equation S_0 and then inserts this into the income equation (I), the intertemporal budget restriction of the households results in Eq. (8).

---

[3] The following analysis is based also on [90].

$$C_1 = S_0(1 + r)$$
$$S_0 = \frac{C_1}{(1 + r)} \tag{8}$$
$$I = I_0 = C_0 + \frac{C_1}{(1 + r)}$$

Thus, the question for the household arises, what is the optimal level of savings for the household, which corresponds to the discounted consumption level of the next period C_1 expressed in Eq. (9).

$$S_0 = \frac{C_1}{(1 + r)} \tag{9}$$

The households have to optimize their benefits according to their preference order received from the consumption today and in the next period, i.e. the household has to weigh between the current and future consumption.

It follows from this that the households have an intertemporal utility function $U = U(C_0, C_1)$ which can be solved by the following Lagrange function (L):

$$L = U(C_0, C_1) + \lambda\left(I - C_0 - \frac{C_1}{1 + r}\right)$$

$$(1) \quad \frac{\partial L}{\partial C_0} = \frac{\partial U}{\partial C_0} - \lambda = 0$$

$$(2) \quad \frac{\partial L}{\partial C_1} = \frac{\partial U}{\partial C_1} - \lambda\frac{1}{1 + r} = 0$$

After conversion we get: $\hspace{4cm}$ (10)

$$(3) \quad \frac{\partial U}{\partial C_0} = \lambda$$

$$(4) \quad \frac{\partial U}{\partial C_1} = \lambda\frac{1}{1 + r}$$

Inserting Eq. (3) in (4), we get:

$$(5) \quad \frac{\partial U}{\partial C_1} = \frac{\partial U}{\partial C_0} \cdot \frac{1}{1 + r}\left|\frac{\partial U}{\partial C_0}\right.$$

and by transforming, based on the first-order conditions we finally receive [90]:

$$\frac{\frac{\partial U}{\partial C_1}}{\frac{\partial U}{\partial C_0}} = \frac{1}{(1 + r)} \tag{11}$$

Equation (11) shows that the intertemporal utility problem can be solved and the utility reaches its maximum, when the marginal utility ratio of future and present consumption corresponds to the discount factor [90].

If, however, the marginal utility ratio is smaller than the discount factor (Eq. 12), it would be more useful for the household to refrain from current consumption and save more in order to be able to consume more in the future.

$$\frac{\frac{\partial U}{\partial C_1}}{\frac{\partial U}{\partial C_0}} < \frac{1}{(1+r)} \tag{12}$$

On the other hand, it is more beneficial for the household to consume more in the present and less in the future, if the marginal utility ratio is greater than the discount rate (Eq. 13).

$$\frac{\frac{\partial U}{\partial C_1}}{\frac{\partial U}{\partial C_0}} > \frac{1}{(1+r)} \tag{13}$$

If the interest rate is zero, then the household would consume the same amount in the present and future.

$$\frac{\partial U}{\partial C_1} = \frac{1}{(1+r)} = \frac{\partial U}{\partial C_0} \tag{14}$$

If the above equation is solved for r, the ratio of time preference rate and interest rate can be shown with Eq. (15):

$$\underbrace{\frac{\frac{\partial U}{\partial C_0} - \frac{\partial U}{\partial C_1}}{\frac{\partial U}{\partial C_1}}}_{\varphi} = r \tag{15}$$

The time preference rate φ in Eq. (15) describes the relative deviation of the marginal utility of the current consumption from the marginal utility of the future consumption [81], i.e. the benefit of the household reaches its maximum, when the time preference rate φ is equal to the interest rate r:

$$\varphi = r \tag{16}$$

Thus, the interest rate corresponds also to the price ratio of the consumed commodities:

$$\frac{P_Q(t)}{P_Q(t+1)} =: 1 + r_t. \tag{17}$$

The ratio indicates how many more units of the consumer goods are obtained in period $t + 1$ if one unit is renounced in period t. With this price ratio, the interest factor is defined [67].

Based on this analysis, it is assumed in the following model (Sect. 4.2) that the time preference rate is equal to the interest rate.

We have outlined the historical background of the current scientific debate about economic growth and its theoretical principles. In the following, we will integrate the FEW nexus, the core sectors of the transition path towards a green economy, in this Solow-based economic growth framework.

3.3 Food–Energy–Water Nexus and Economic Growth

Food, energy and water are crucial resources and key elements of life and of any human welfare. The nexus among the key elements means that water, energy and food production are closely connected [3]. This food–energy–water nexus builds an "intrinsic link [10]" between the global challenges of ensuring food security and the rising demand for energy and water, as the UN determined in 2014 [82, 84]. These central resources for any society are under increasing stress:

- The OECD assumes that the global water demand will increase by 55% by 2050, so that 40% of the global population will be living under water stress conditions [54].
- The water demand for energy production will increase by about 20% [35].
- 800 million people are currently hungry and by 2050 food production will have to increase by about 50% due to the rising world population [24].
- And climate change will further set the water resources under additional stress because of higher temperatures, resulting in "shifts in precipitation patterns and snow cover, and a likely increase in the frequency of flooding and droughts [23]", which will also have an impact on the global food production.

These key sectors and their interrelated systems are under severe stress yet central for any life on earth. Their governance must consider socio-economic, ecological and governmental indicators irrespective of the chosen growth scenarios in the decisions. Figure 6 shows that the FEW nexus is a triple-encircled system. Accordingly, a well-balanced management of the FEW nexus is dependent (a) on the surrounding economic system characterized by key economic indicators, (b) on environmental indicators developed by Raworth [63] and (c) on characteristics of good governance developed by the UNDP [79, 80]. These indicators define the solution space for the management of the FEW nexus, as Fig. 6 shows.

The nine Raworth indicators [63] (climate change, land use change, biodiversity loss, ozone depletion, chemical pollution, nitrogen phosphorus cycles, ocean acidification, atmospheric aerosol loading, freshwater use) describe the environmental system of the FEW nexus and its constraints. The described environmental system thus provides the resources for the economic system, building thereby the natural constraints for the economic system that affect also the FW nexus. On the other hand, the economic system is restricted by the Raworth indicators and can be characterized by nine indicators of the presented economic growth model of Chap. 4 (capital,

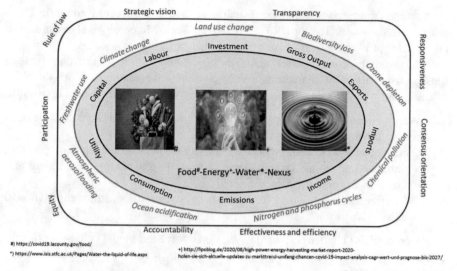

Fig. 6 The solution space of the FEW nexus—indicators and constraints. *Source* Authors (2020) based on Raworth [63] and UNDP [79, 80]

labour, investment, gross output, exports, imports, income, emissions, consumption, and the utility level of society), as Fig. 6 reveals. The development of the capital expenditures and of the labour demand is determining the gross output of the sector, whereas the investments define the future prosperity of the sectors, the trade relations define how far the nexus sector is embedded in the current globalized economy. The income and consumption determine to some extent the utility level of the households. Furthermore, the management of the FEW nexus should be based on the characteristics of good governance (participation, rule of law, transparency, responsiveness, consensus orientation, equity, effectiveness and efficiency, accountability, and strategic vision) developed by the UNDP [78, 79]. A good governance-based management should consider in its decisions both the results of the environmental and of the economic indicators.

Figure 6 shows how closely interwoven the FEW nexus is with the economic, political and ecological systems, and how dependent the nexus sectors are on all three surrounding systems. The management of the FEW nexus based on good governance has to consider these economic and ecological constraints in the concept. Therefore, in our model the interaction between the FEW nexus and the economic–ecological system will be analysed delivering data and information for the management process of the FEW nexus.

Based on the historical analysis and the current economic developments before and during the Corona pandemic [59], the idea of Porter's thought experiment in the New York Times is taken up. The paper discusses leaving the current global economic model for some countries behind, imagining instead a world with and without economic growth to analyse if such an economic world could reduce the CO_2-emissions as demanded by the IPCC [36] and the World Meteorological Organization (WMO) [87] to limit global warming to 1.5 °C [36].

Addressing this controversial debate, the objective of our model is to contribute to the discussion about the characteristics of a post-growth economy in which the FEW nexus is embedded. Research has to address a number of growth-related questions:

1. Which effect will different growth scenarios have on emissions arising from production and consumption?
2. Can an economic system be organized as a zero or de-growth economy within a globalized economic growth system? And, if so, what are the national and international economic consequences?
3. What effects will such a transformation have on occupation, utility, capital needs, and the FEW nexus?

Based on these challenges, we derived a new, stylized model consisting of four countries to discuss the effects of different growth scenarios on the various key economic indicators.

4   A Four Country Dynamic Multinational CGE Model

Solow argues that his growth model is based on theoretical assumptions because "all theory depends on assumptions which are not quite true. That is what makes it theory. The art of successful theorizing is to make the inevitable simplifying assumptions [72]" based on a realistic view on the economic development.

Hence, we start our analysis with the definition of our assumptions to approach reality.

4.1   Introduction

We use a dynamic multinational general equilibrium model (GE model) based on ECOMOD 2003. Similar models were used, for example, also by Bretschger [13]. The modelling approach was inspired by Kaldor [39] and is in line with the tradition of the Arrow–Debreu model [7, 8], the latter of which was used to prove the existence of equilibria [33].

There is no uncertainty or no money illusion. We assume that all economic actors have infinite horizons with rational forward-looking expectations [53, 65, 66] and, thus, fully dynamically optimize their decision variables for $t = 12$ periods. Those economic agents include the industry that consists of various sectors of production, households and governments for each of the four countries considered.

The model equations that describe the optimal behaviour of all agents are explained in the next subsection, while the database and calibration of relevant parameters will be discussed afterwards. The database does not describe the economic situation of a real country. Instead, we chose a fictive database for two reasons:

- First, this paper is not a forecast for a particular country but aims a revealing essential mechanisms correlated to different growth scenarios.
- Second, the stylized model approach was chosen to reveal the effects of the different growth strategies without disturbing side effects such as the Lehman crash 2008, the Euro crises 2012, or the current Corona caused economic shutdowns 2020.

4.2 The Model

In each country $Co = A, B, C, D$ a representative consumer behaviour can be characterized by an intertemporal utility function U_{Co} [17, 21]

$$U_{Co} = \sum_{t=1}^{\infty} \left(\frac{1}{1 + \varphi_{Co}} \right)^t \ln U_t,$$

φ = time preference rate, Co = Countries A, B, C, D, i = sector (18)

The consumer maximizes its utility for time periods $t = 1, \ldots, 12$ subject to the budget constraints

$$U_{t,Co} = \prod_{i=1}^{3} C_{t,i,Co}^{\alpha Hi}, \alpha_{H,Co,i}$$

α are the share parameters in the Cobb-Douglas
utility function whose sum is equal to 1 (19)

The available budget in period t results from income $Y_{t,Co}$ plus government transfers $TRF_{t,Co}$ and dividends $DIV_{t,Co}$ can be spent on consumption $C_{t,i,Co}$ and savings $S_{t,Co}$.

From this maximization problem, optimal savings and per period consumption can be derived. Consumption is determined for each of the industry sectors.

Each sector of our model is represented by a firm that chooses, at each period, the input levels of labour and makes investment decisions to maximize the value of the firm. They use a constant returns to scale Cobb–Douglas technology, where $K_{t,i,Co}$ denotes capital in period t, $L_{t,i,Co}$ is the labour input and $I_{t,i,Co}$ denotes investment. Domestically produced output $XD_{t,i,Co}$ thus results from the following production function:

$$\text{XD}_{t,i,\text{Co}} = f_{t,i}\big(K_{t,i}L_{t,i}\big) = \alpha K_{t,i,\text{Co}}^{\text{VKS}_{i,\text{CO}}} \cdot L_{t,i,\text{Co}}^{1-\text{VKS}_{i,\text{CO}}}, \quad \text{Co} = \text{Countries } A, \ B, \ C, \ D \tag{20}$$

$\text{XD}_{t,i,\text{Co}}$ will be used for domestic supply $\text{XDD}_{t,i,\text{Co}}$ and exports $\text{Ex}_{t,i,\text{Co}}$. Exports are determined in the market clearance condition

$$\text{XD}_{t,i,\text{Co}} - \phi_{i,\text{Co}} \cdot I_{t,i,\text{Co}}^{\left(\frac{2}{K_{t,i,\text{Co}}}\right)} = \text{XDD}_{t,i,\text{Co}} + E_{t,i,\text{Co}} \tag{21}$$

so that exports plus domestically produced goods aimed at domestic supplies $\text{XDD}_{t,i,\text{Co}}$ absorb the total quantity of domestically produced goods $\text{XD}_{t,i,\text{Co}}$ which is left after accounting for investment $I_{t,i,\text{Co}}$. Investment is calculated to compensate depreciation.

Our model includes four countries that depend on each other via their trade relations. For imports versus domestically produced goods, we apply the Armington assumption [5, 6]. The basic idea of the Armington assumption is that domestically produced output and imports can be considered imperfect substitutes from the consumer's point of view, accounting for product heterogeneity [11, 28]. Therefore, domestically produced goods aimed at domestic supply $\text{XDD}_{t,i,\text{Co}}$ and imports $\text{IM}_{t,i,\text{Co}}$ can be combined in the aggregate quantity $X_{t,i,\text{Co}}$ using a CES production function. The Armington assumption is commonly used in CGE models [43, 49].

For the sake of simplicity, we assume that import prices $\text{PM}_{t,i,\text{Co}}$ equal the exporting countries local prices for the respective goods. For example, if country A imports from country C, A pays an import price $\text{PM}_{t,i,A,\text{Co}}$ equal to C's local price $\text{PM}_{t,i,\text{Co}}$:

$$\text{PM}_{t,i,A,\text{Co}} = \text{PD}_{t,i,\text{Co}} \tag{22}$$

The balance of payment depends on imports and exports only and is defined as:

$$\sum_{\text{Co}=1}^{4} \sum_{i=1}^{3} \text{PM}_{t,i,\text{Co},\text{Co}} \cdot M_{t,i,\text{Co},\text{Co}} = \sum_{i=1}^{3} \text{PD}_{t,i,\text{Co}} \cdot E_{t,i,\text{Co}} \tag{23}$$

The government collects taxes from emissions and redistributes them in terms of household transfers. The government's budget is assumed to be in balance.

$$\text{TRF}(t, \text{Co}) = \text{sum}(i, \text{Taxr}_{i,t,\text{Co}}) \tag{24}$$

Our computable general equilibrium (CGE) model has the following standard characteristics: As demand is homogenous of degree zero in the price vector, only relative prices are determined. Following Walras' Law (which determines that if $n - 1$ markets are in equilibrium, the nth market must be in equilibrium as well), we omit one of the market clearance conditions [69].

4.3 Calibration

Our multiregion CGE model requires a consistent database that describes the economic status quo of the four countries. The database is necessary for the calibration of the missing parameters. For our four countries, we developed the Social Accounting Matrices (SAM) [64] accounting for the three economic sectors:

Sector 1: Food–Energy–Water (FEW)

Sector 2: Industry production

Sector 3: Public and private services.

The social accounting matrix is a structured representation of all relevant data. This includes consumer demand or the use of private income for example. The database must be consistent in the sense that the consumer spends its income on consumption and savings without leaving money on the table according to our model equations. This applies to values and quantities for all our economic agents.

We further assume that the initial data of country A is identical to the data set of country D, and likewise countries B and C are initially identical, see Tables 6 and 7.

Table 6 Social accounting matrix countries A and D

Country $A + D$, initial data	FEW	Industry	Service	Sum
Consumer demand	70	120	185	375
Investment	5	40	100	145
Capital stock	50	60	100	210
Labor demand	20	90	200	310
Export	15	20	50	85
Import	20	30	35	85

Source Authors (2020) based on ECOMOD [21] and IEK-STE/SRH (2020)

Table 7 Social accounting matrix countries B and C

Country $B + C$, initial data	FEW	Industry	Service	Sum
Consumer demand	125	210	415	750
Investment	250	420	200	870
Capital stock	20	160	400	580
Labor demand	200	240	400	840
Export	20	360	35	415
Import	15	80	50	145

Source Authors (2020) based on ECOMOD [21] and IEK-STE/SRH (2020)

Table 8 Key model parameters

	Country			
	A	B	C	D
Interest rate (%)	5	5	5	5
Time preference rate (%)	5	5	5	5
Steady-state growth rate (%)	−0.001	1.2	1.9	−1.3
Labour development (%)	0.001	0.001	0.001	0.001

Source Authors (2020) and IEK-STE/SRH (2020)

The social accounting matrices allow for the calibration of model parameters, such as the share parameter in the production function. Furthermore, some assumptions have to be made for parameters that cannot be calculated. Those include the steady-state growth rates, the interest rate or time preference rates. Table 8 gives an overview of the key assumptions.

As we want to elaborate specifically and exclusively on the effects of different growth rates, we vary those between the countries. The interest rate equals the time preference rate. This allows for a ceteris paribus comparison of the results as differences can be traced back to the difference in growth rates. Particularly, the growth rates are:

- Country A is in a zero-growth scenario based on the ideas of Maxton [50] and Jackson [38],
- Country B will grow according to the Randers model [62] by 1.2% per year [62],
- Country C will grow conventionally by 1.9%,
- Country D will decrease by 1.3%, based on the ideas and models developed by Victor [86], Weitzman [89] and Paech [57, 58].

5 Model Results

In the following, the results of our stylized economic model for the four countries will be presented. The impact of our four scenarios and its economic framework conditions on the FEW nexus sector and on the other economic indicators (i.e. income, consumption, savings, investments, gross output, trade relations, utility level and the emissions of the countries) will be shown.

Table 9 shows that the income of the four countries reacts differently based on the respective country growth scenario. For the countries A and D with their negative and zero-growth rate assumptions, the model calculations reveal that the income of country A decreases from 506 monetary units to 500 over the observed 12-year period. On the other hand, the income of country B rises from 1011 monetary units to 1150 at the end of the period, and the income of country C also increases from 1012 to 1243, whereas the income of country D in the de-growth scenario decreases

Table 9 Development of the income of the four countries—in nominal monetary units

Country (growth rate)/year	1	2	3	4	5	6	7	8	9	10	11	12
A (−0.0%)	506	505	504	504	504	503	503	502	502	501	501	500
B (1.2%)	1011	1023	1035	1048	1049	1070	1083	1096	1110	1123	1137	1150
C (1.9%)	1012	1031	1053	1075	1078	1111	1131	1153	1175	1197	1220	1243
D (−1.3%)	506	500	494	487	480	475	468	462	456	451	444	439
Sum of income	3035	3059	3086	3114	3112	3158	3186	3214	3243	3272	3301	3331

Source Own calculations (2020) and IEK-STE/SRH (2020)

from 506 to 439 monetary units. The de-growth and zero-growth strategy results in a significant reduction of the income of the two countries.

The development of the income has a direct impact on the consumption of the four countries, especially on the commodities of the FEW sector. Table 10 shows that the consumption of the four countries based on the earned income develops in different ways.

The consumption of the food–energy–water sector goods in country A stays more or less the same during the 12-year period, whereas the consumption of these goods in countries B and C increases from 118 monetary units to 135 and 145 monetary units respectively, so that the consumption of all produced FEW goods increased from 369 to 403 monetary units. The consumption of the industry and service goods of country A decreased slightly, whereas the consumption of these goods increased for the country B and C and is reduced clearly for the de-growth country D.

The not used part of the income is saved for later consumption or investments in the four countries and this pattern is also reflected in the development of the country savings (Table 11). The savings of the countries B and C increase over the analysed

Table 10 Development of the consumption of FEW sector commodities of the four countries—in nominal monetary units

Country (growth rate)/year	1	2	3	4	5	6	7	8	9	10	11	12
Food–energy–water—A (−0.0%)	67	66	66	66	66	66	66	66	66	66	66	66
Industry and service—A	294	293	293	293	293	292	292	292	292	291	291	291
Food–energy–water—B (1.2%)	118	120	121	123	124	125	127	128	130	131	133	134
Industry and service—B	604	611	618	626	634	639	646	654	662	670	678	686
Food–energy–water—C (1.9%)	118	120	123	125	128	130	132	135	137	140	143	145
Industry and service—C	603	615	627	640	653	662	675	687	700	714	727	742
Food–energy–water—D (−1.3%)	67	66	65	64	63	62	62	61	60	59	58	58
Industry and service—D	294	291	287	283	279	276	272	269	265	262	258	255

Source Own calculations (2020) and IEK-STE/SRH (2020)

Table 11 Savings of the single countries and the total savings—in nominal monetary units

Country (growth rate)/year	1	2	3	4	5	6	7	8	9	10	11	12
A (−0.0%)	144.5	145.2	144.8	144.6	145.0	144.3	144.2	144.0	143.8	143.7	143.6	143.5
B (1.2%)	289.2	292.4	295.9	299.6	291.9	305.9	310.4	314.2	318.0	321.9	325.9	329.0
C (1.9%)	290.4	296.1	303.2	309.9	297.1	319.7	324.9	331.0	337.6	343.7	350.1	356.2
D (−1.3%)	145.4	143.6	141.9	140.4	137.8	136.1	134.4	132.6	131.0	129.4	126.7	125.8
Sum of savings	869.5	877.3	885.8	894.4	871.8	906.0	913.9	921.8	930.4	938.8	946.2	954.5

Source Own calculations (2020) and IEK-STE/SRH (2020)

Table 12 Development of the investments in FEW sector and the other two sectors—in nominal monetary units

Country (growth rate)/year	1	2	3	4	5	6	7	8	9	10	11	12
Food–energy–water—A (−0.0%)	34	35	34	34	34	34	34	34	34	34	34	34
Industry and service—A	110	111	110	110	110	110	110	110	110	110	109	109
Food–energy–water—B (1.2%)	68	69	69	70	69	72	73	74	74	75	76	77
Industry and service—B	221	224	226	229	224	234	238	241	244	246	249	252
Food–energy–water—C (1.9%)	69	70	72	74	71	76	77	79	80	82	83	85
Industry and service—C	221	226	231	236	226	244	248	252	257	262	267	271
Food–energy–water—D (−1.3%)	69	69	69	69	69	69	69	69	69	69	68	68
Industry and service—D	111	109	108	106	105	104	102	101	100	99	96	96

Source Own calculations (2020) and IEK-STE/SRH (2020)

12-year period, whereas the country D faces a significant decline in its savings, whereas it remains unchanged for country A.

The developments of the investments, which are based on the savings and define the future prosperity perspectives of the four countries, also differ along the different growth strategies, as Table 12 shows.

The investments grow only in countries B and C, whereas the other two countries (A and D) show remaining or declining investments in future prosperity. The investments of countries B and C in the FEW sector increase from 68/69 to 77 and 85 monetary units, whereas the level of FEW investments of country A is constant over the observed period, but investments decrease country D. The investments decline for both the FEW sector and the other two economic sectors of region D.

Table 13 focuses on the gross output (output approach) of the four economies. In countries A and D with negative growth rates, the gross output decreases by also negative rates, whereas the gross output of countries B and C increases slightly but less than the initial growth rate of the countries.

The gross output of the FEW nexus sector of country A remains more or less the same, whereas the output of this sector in country D declines continuously in the observed period. The FEW nexus of country B and C increases from 140 to 157 and in the case of country B to 172. The output of the industry and service sectors of country A declines from 450 to 445 and in case of country D to 389, whereas these sectors' outputs increase in countries B and C from 900 to 1027 and 1106.

The gross output of the four countries is produced by the use of the production factors labour and capital, whose development is shown in Tables 14 and 15. In countries A and D, the labour demand decreases from 310 to 307 in case of country A and to 269 for country D. Labour demand of the countries B and C increases from 620/621 labour units to 706 and 762 labour units at the end of the observed time. Total labour demand of all four countries increases from 1861 to 2044.

Table 15 reveals the capital requirements, which are needed to employ the labour force presented in Table 14. The capital requirements of the countries A and D decrease from 825 and 668 monetary units to 817 and 578 monetary units, whereas

Table 13 Development of the gross output of the FEW nexus sector goods and the other two sectors—in real good units

Country (growth rate)/year	1	2	3	4	5	6	7	8	9	10	11	12
Food–energy–water—A (−0.0%)	70	70	70	70	70	70	70	70	69	69	69	69
Industry and service—A	450	449	449	449	449	448	448	447	447	446	446	445
Food–energy–water—B (1.2%)	140	141	143	144	146	147	148	150	152	153	155	157
Industry and service—B	900	911	922	933	945	954	966	978	990	1002	1015	1027
Food–energy–water—C (1.9%)	140	143	145	148	151	153	156	159	162	166	169	172
Industry and service—C	900	917	935	953	971	988	1006	1026	1045	1065	1086	1106
Food–energy–water—D (−1.3%)	70	69	68	68	68	66	65	64	63	62	62	61
Industry and service—D	450	444	438	433	424	421	416	411	405	400	395	389

Source Own calculations (2020) and IEK-STE/SRH (2020)

Table 14 Development of labour input of the four countries—nominal wage

Country (growth rate)/year	1	2	3	4	5	6	7	8	9	10	11	12
Country A (−0.0%)	310	310	309	309	309	309	308	308	308	307	307	307
Country B (1.2%)	620	627	635	643	643	656	665	673	681	690	699	706
Country C (1.9%)	621	632	646	659	659	682	694	707	721	734	748	762
Country D (−1.3%)	311	307	303	299	294	291	287	283	280	276	272	269
Sum of labor input	1861	1876	1893	1910	1906	1937	1954	1971	1990	2008	2026	2044

Source Own calculations (2020) and IEK-STE/SRH (2020)

Table 15 Development of capital input

Country (growth rate)/year	1	2	3	4	5	6	7	8	9	10	11	12
A (−0.0%)	825	824	823	823	821	822	821	820	819	818	817	817
B (1.2%)	2215	2241	2267	2293	2294	2337	2367	2396	2425	2455	2485	2515
C (1.9%)	2718	2770	2828	2889	2889	2980	3035	3094	3153	3214	3275	3335
D (−1.3%)	669	661	652	645	637	627	619	611	603	596	586	578
Sum of capital input	6427	6495	6570	6649	6642	6766	6842	6921	7001	7083	7164	7245

Source Own calculations (2020) and IEK-STE/SRH (2020)

the capital requirements of the countries B and C increase constantly to 2515 and 3335 monetary units.

The analysis of our model reveals further the impact of the various growth scenarios on the trade relations of the four countries. The share of country A on the 4-country trade remains more or less the same over the observed period, whereas the share of country B increases by about 14% and of country C by about 23%. The share of the de-growth country D on the total trade decreases by about 14%, as Table 16 shows.

The 4-country model determined also the emissions of the four countries over the observed time, as Table 17 shows. The indicator "emissions" includes all emissions, which occur in the production and consumption process.

Emissions decline constantly for the two countries (A, D), whereas the emissions of country B increase annually by about 1.2% and of country C by about 1.8%. In country B, the increase of the emissions is strongly correlated to the growth

Table 16 Development of the trade relations—share of single country on 4-country trade. $1 = 100$

Country (growth rate)/year	1	2	3	4	5	6	7	8	9	10	11	12
A (−0.0%)	100	100	100	99	106	100	99	99	99	99	99	99
B (1.2%)	100	101	102	104	103	106	107	109	110	111	113	114
C (1.9%)	100	102	104	106	110	110	112	114	116	118	121	123
D (−1.3%)	100	99	97	96	91	93	92	91	90	89	88	86
Sum of total trade	100	100	101	101	103	102	103	103	104	104	105	106

Source Own calculations (2020) and IEK-STE/SRH (2020)

Table 17 Development of production- and consumption-based emissions of the four countries

Country (growth rate)/year	1	2	3	4	5	6	7	8	9	10	11	12
A (−0.0%)	106	105	105	105	105	105	105	105	105	105	105	105
B (1.2%)	194	196	198	201	204	205	208	210	213	215	218	221
C (1.9%)	176	179	183	186	191	193	197	206	204	208	212	216
D (−1.3%)	123	122	120	119	116	115	114	112	111	110	108	107
Sum of emissions	599	603	607	611	617	619	624	634	633	638	643	648

Source Own calculations (2020) and IEK-STE/SRH (2020)

rate, whereas for country D the emission reduction is stronger than the growth rate reduction. Despite the de-growth and zero-growth approaches of countries A and D, the total emissions of the 4-country economy grow from 599 to 648 emission quantity units.

The growth rate scenarios of the four countries also show an impact on the utility of the households of the countries (Table 18).

The first results of our model show that the utility development largely correlates to the development of the growth rate of the country. In country A, the utility level declines by -0.1% per annum, whereas the utility levels of the countries B and C increase by 1.0 and 1.74% per year respectively, while the utility level of the de-growth country D faces a more significant decline in the utility level of the households.

6   Outlook

The historical background analysis of our thought experiment has shown that economic growth is not a natural development but rather a by-product of the industrialization in Western Europe and its Offshoots in the nineteenth century. The development of China has shown that a region can live for nearly 2000 years with almost no per capita economic growth.

Table 18 Utility

Country (growth rate)/year	1	2	3	4	5	6	7	8	9	10	11	12
A (−0.0%)	129.5	129.0	129.0	128.9	129.1	128.8	128.6	128.5	128.4	128.3	128.1	128.0
B (1.2%)	271.3	274.6	277.8	281.1	287.9	287.2	290.4	294.0	297.5	301.1	304.8	308.8
C (1.9%)	270.8	275.9	230.8	286.1	297.7	296.8	302.7	308.6	314.4	320.5	326.7	333.1
D (−1.3%)	129.3	127.6	125.9	124.3	121.7	121.2	119.6	118.0	116.5	114.9	113.7	111.9

Source Own calculations (2020) and IEK-STE/SRH (2020)

The analysis also shows how the FEW nexus sector develops over the observed time. In the zero and de-growth scenarios, gross output and consumption of the nexus products remain the same over the observed period in the zero-growth (A) scenario, and in the de-growth scenario (D) the demand for the nexus commodities declines, whereas in the other two countries (B, C) the demand increases continuously. The decline and unchanged investments in the nexus sectors of country A and D will have a negative impact on the prosperity and adaptability of its FEW nexus sectors, whereas in the other countries B and C the investments increase in the 12-year period and build the foundation of future economic growth with increasing resource consumption and rising emissions.

Our four-country-stylized GE model further reveals the economic effects of a zero-growth and de-growth strategy. Overall, the model demonstrates that restrictions on economic growth have repercussions on all economic variables. The zero-growth and de-growth strategies confirm the hypothesis that these strategies reduce the production and consumption emissions.

The loss of economic growth leads to a reduction in the utility of the households and a reduction of labour and capital demand and of future prosperity due to the decline of investment. The question arises how these losses can be offset by other societal, political and institutional developments in the countries with a de-growth or zero-growth approach. These alternatives would have to provide new job and utility generating opportunities to compensate the observed utility losses and enable a sustained, inclusive and sustainable economic growth to avoid distortions as currently imposed by the Corona pandemic. New institutional and political measures would have to account for the fact that "in a world economy that does not grow, the powerless and vulnerable are the most likely to lose [59]". Further research is needed on how the decline in utility of the households of these countries can be balanced economically and politically. This decision-making process should involve all stakeholders and institutions, so that good governance has to mediate the differing interests in the socio-economic adaptation process. Additional research is also needed on the economic effects of relocating the workforce and the capital stock from formal classical economic sectors to a new de-growth economic world, in which the FEW nexus will build the centre.

7   Recommendations for Further Research

The previous analysis has shown that in the case of negative or zero-growth scenarios, a decline of the labour demand in countries A and D (Table 14) can be observed. The following research question thus arises: Can these countries develop a "de-growth compensation sector" to compensate for and take up the labour force released by the traditional economic sector? In this context, it should also to be analysed, if this new alternative sector can even out the utility reduction (see Table 18) caused by the growth strategy in the traditional economic sectors. It has to be investigated, if the new "de-growth compensation sector" can provide enough new job opportunities especially for the losses in the FEW nexus sectors to compensate households for their utility loss.

Furthermore, it has to be examined, if this "de-growth compensation sector" can be a role model for the ideas of 'networks of neighbourhood help' developed by Jackson and Paech [37, 58]. To analyse their new prosperity ideas, a derived "de-growth compensation sector" sector has to be integrated in the CGE model. The GE model thus enables us to derive the labour and capital needs of the new de-growth sector and analyse if this "de-growth compensation sector" reduces the CO_2-emission through a change of lifestyle, institutional innovations, or neighbourhood help.

Further institutional research is needed to prove, if the "de-growth compensation sector" sector can be seen as a first institutional step towards a world without economic growth and if the transformation of the economic system also needs a new set of institutional incentives [9] especially in the core of sustainable development—i.e. the FEW nexus sectors—to protect the socio-economic-ecological system.

Hence, a further question has to be discussed: Is the FEW nexus sector in the future a divided sector? One classical part in the traditional economy and a new part in the "de-growth compensation sector" sector. Additionally, the research has to be widened to the question about the impact of zero and de-growth scenarios on international trade. These research questions have to be included also in an extended CGE trade model.

References

1. Acemoglu D (2009) Introduction to modern economic growth. Princeton University Press, Princeton
2. Alfaro L, Jeong S (2020) COVID-19: the global shutdown. In: Harvard Business School case, pp 320–108
3. Allouche J, Middleton C, Gyawa D (2019) The water–food–energy nexus: power, politics, and justice. Routledge, New York
4. Amigo I (2020) When will the Amazon hit a tipping point? Nature 578:505–507. https://doi.org/10.1038/d41586-020-00508-4
5. Armington PS (1969) A theory of demand for products distinguished by place of production. Staff Pap (Int Monetary Fund) 16:159–178
6. Armington PS (1990) Endogenous limits of growth—politicoeconomic analysis of distribution conflicts in democratic systems. Polit Vierteljahr 31:706–709
7. Arrow KJ (1974) General economic equilibrium: purpose, analytic techniques, collective choice. Am Econ Rev 64:253–272
8. Arrow KJ, Debreu G (1954) Existence of an equilibrium for a competitive economy. Econometrica 22:265–290. https://doi.org/10.2307/1907353
9. Arrow K et al (1995) Economic growth, carrying capacity, and the environment. Science 268:520–521
10. Beddington J (2009) Food, energy, water and the climate: a perfect storm of global events? Chief Scientific Adviser to HM Government, UK Government Office for Science, London
11. Bosello F, Roson R, Tol R (2007) Economy-wide estimates of the implications of climate change: sea level rise. Environ Resour Econ 37:549–571

12. Boulding K (1994 [1966]) The economics of the coming spaceship earth. In: Daly H, Townsend K (eds) Valuing the earth, economics, ecology, ethics. MIT Press, Boston, pp 297–310
13. Bretschger L, Ramer R, Schwark F (2011) Growth effects of carbon policies: applying a fully dynamic CGE model with heterogeneous capital. Resour Energy Econ 33:963–980. https://doi.org/10.1016/j.reseneeco.2011.06.004
14. Böhm-Bawerk E (1961a) Geschichte und Kritik der Kapitalzins-Theorien, 4th edn. Verlag Anton Hain K.G., Meisenheim/Glan
15. Böhm-Bawerk E (1961b) Positive Theorie des Kapitales vol Erster Band, Buch I–IV, 4th edn. Verlag Anton Hain K.G., Meisenheim/Glan
16. Carson R (2002 [1962]) Silent spring, 40th anniversary edn. Houghton Mifflin Company, Boston
17. Chiang AC (1984) Fundamental methods of mathematical economics. In: Economics series. McGraw Hill, Singapore
18. Daly HE (1974) The economics of the steady state. Am Econ Rev 64:15–21
19. Domar ED (1946) Capital expansion, rate of growth, and employment. Econometrica 14:137–147. https://doi.org/10.2307/1905364
20. D'Amato D et al (2017) Green, circular, bio economy: a comparative analysis of sustainability avenues. J Clean Prod 168:716–734. https://doi.org/10.1016/j.jclepro.2017.09.053
21. ECOMOD (2003) Practical general equilibrium modeling using GAMS. EcoMod Press, Northampton
22. European Environment Agency (2019) The European environment—state and outlook 2020. Knowledge for transition to a sustainable Europe. EEA, Denmark
23. European Environment Agency (EEA) (2018) Climate impacts on water resources. EEA. https://www.eea.europa.eu/archived/archived-content-water-topic/water-resources/climate-impacts-on-water-resources. Accessed 2 Oct 2020
24. FAO, IFAD, UNICEF, WFP, WHO (2017) The state of food security and nutrition in the world 2017. Building resilience for peace and food security. FAO, Rome
25. Faber M (ed) (1986) Studies in Austrian capital theory, investment and time. Springer Heidelberg, Berlin
26. Faber M (1987) Introduction to modern Austrian capital theory. Springer Heidelberg, New York
27. Food and Agriculture Organization (, FAO,) (2014) The water-food-energy nexus: a new approach in support of food security and sustainable agriculture. Food and Agriculture Organization of the United Nations, Rome
28. Francois JF, Reinert KA (eds) (1997) Applied methods for trade policy analysis. Cambridge University Press, Cambridge
29. Georgescu-Roegen N (1971) The entropy law and the economic process. Harvard University Press, Cambridge
30. Gutteres A (2019) Davos speech 2019, 24 Jan 2019. World Economic Forum, Davos
31. Hardin GJ (1968) The tragedy of the commons. Science 162
32. Harrod RF (1939) An essay in dynamic theory. Econ J 49:14–33. https://doi.org/10.2307/2225181
33. Hennipman (2008) General equilibrium. Macmillan Publishers, New York
34. Hoff H (2011) Understanding the nexus—background paper for the Bonn2011 conference: the water, energy and food security nexus. Stockholm Environment Institute (SEI), Stockholm
35. IEA (2012) World energy outlook 2012. IEA, Paris
36. IPCC (2018) Global warming of 1.5°C—summary for policymakers. IPCC, Incheon
37. Jackson T (2009) Prosperity without growth. Earthscan, London
38. Jackson T (2019) The post-growth challenge: secular stagnation, inequality and the limits to growth. Ecol Econ 156:236–246. https://doi.org/10.1016/j.ecolecon.2018.10.010
39. Kaldor N (1957) A model of economic growth. Econ J 67. https://doi.org/10.2307/2227704
40. King MD et al (2020) Dynamic ice loss from the Greenland ice sheet driven by sustained glacier retreat. Commun Earth Environ 1:1. https://doi.org/10.1038/s43247-020-0001-2
41. Lange S (2018) Macroeconomics without growth. Metropolis-Verlag, Marburg

42. Le Quéré C et al (2020) Temporary reduction in daily global CO_2 emissions during the COVID-19 forced confinement. Nat Clim Change. https://doi.org/10.1038/s41558-020-0797-x
43. Lofgren H, Cicowiez M (2018) Linking Armington and CET elasticities of substitution and transformation to price elasticities of import demand and export supply: a note for CGE practitioners. Documento de Trabajo, Universidad Nacional de La Plata
44. Maddison A (2001) The world economy: a millennial perspective. OECD Development Centre, Paris
45. Maddison A (2008) The west and the rest in the world economy: 1000–2030. Maddisonian and Malthusian interpretations. World Econ 9:75–100
46. Maddison A (1995) Monitoring the world economy, 1820/1992. Development Centre Studies, OECD Publishing, Paris
47. Maddison A (2007) Contours of the world economy, 1–2030 AD, essays in macro-economic history. Oxford University Press, Oxford
48. Maddison A (Groningen Growth and Development Centre (GGDC)) (2009) Historical statistics in www.ggdc.net/Maddison. Groningen Growth and Development Centre (GGDC). https://www.rug.nl/ggdc/historicaldevelopment/maddison/original-maddison
49. Martin WJ (1997) Measuring welfare changes with distortions. In: Francois JF, Reinert KA (eds) Applied methods for trade policy analysis. Cambridge University Press, Cambridge
50. Maxton G (2018) Change! Komplett Media, München
51. Meadows DH, Meadows DL, Randers J, Behrens WW (1972) The limits to growth. Universe Books, New York
52. Mega ER (2020) 'Apocalyptic' fires are ravaging the world's largest tropical wetland. Nature 586. https://doi.org/10.1038/d41586-020-02716-4
53. Morris S, Shin HS (2006) Information diffusion in macroeconomics—inertia of forward-looking expectations. MIT paper
54. OECD (2012) Environmental outlook 2012. OECD, Paris
55. OECD (2020) Coronavirus (COVID-19): living with uncertainty. OECD, Paris
56. Ostrom E (1990) Governing the commons. The evolution of institutions for collective action. Cambridge University Press, Cambridge/New York/Victoria
57. Paech N (2018) Befreiung vom Überfluss. Oekom, München
58. Paech N (2020) Postwachstumsökonomik. Gabler Lexikon. https://wirtschaftslexikon.gabler.de/definition/postwachstumsoekonomik-53487. Accessed 11 May 2020
59. Porter E (2015) Imagining a world without growth. New York Times, 1 Dec 2015, New York
60. Post-Growth 2018 Conference (2018) Post-growth open letter to EU institutions signed by over 200 scientists: "Europe, it's time to end the growth dependency". Post-Growth 2018 Conference. https://degrowth.org/2018/09/06/post-growth-open-letter/. Accessed 20 Sept 2020
61. Rampell C (2010) Angus Maddison, economic historian, dies at 83. New York Times, New York
62. Randers J, Maxton G (2016) Ein Prozent ist genug. Oekom Verlag, München
63. Raworth K (2012) A safe and just space for humanity. Can we live within the doughnut? Oxfam discussion papers 2012
64. Roland-Holst DW (1997) Social accounting matrices. In: Francois JF, Reinert KA (eds) Applied methods for trade policy analysis. Cambridge University Press, Cambridge
65. Sargent TJ (1987) Rational expectations
66. Savin NE (1987) Rational expectations: econometric implications
67. Schmutzler A (1991) Grundlagen der Neo-Österreichischen Kapitaltheorie. Heidelberg
68. Schumacher EF (1973) Small is beautiful: a study of economics as if people mattered. Blond and Briggs, London
69. Shoven JB, Whalley J (1993) Applying general equilibrium. Cambridge University Press, Cambridge
70. Smil V (1991) General energetics. John Wiley & Sons, New York
71. Smil V (1994) Energy in world history. Westview, Boulder
72. Solow RM (1956) A contribution to the theory of economic growth. Q J Econ 70:65–94. https://doi.org/10.2307/1884513

73. Stiglitz JE, Sen A, Fitoussi J-P (2010) Mismeasuring our lives: why GDP doesn't add up. New Press, New York
74. Stiglitz JE, Sen A, Fitoussi J-P (2009) Report by the commission on the measurement of economic performance and social progress. Paris
75. Swan TW (1956) Economic growth and capital accumulation. Econ Rec 32:334–361. https://doi.org/10.1111/j.1475-4932.1956.tb00434.x
76. The Economist (2010) Economics focus: Maddison counting, vol 2010. London
77. Trainer T (2020) De-growth: some suggestions from the simpler way perspective. Ecol Econ 167:106436. https://doi.org/10.1016/j.ecolecon.2019.106436
78. U.S. Bureau of Economic Analysis (BEA) (2020) News release: gross domestic product, second quarter 2020. BEA, Washington
79. UNDP (1997) Characteristics of good governance. Global Development Research Centre. https://www.gdrc.org/u-gov/g-attributes.html
80. UNDP (2014) Governance for sustainable development. Discussion paper 14
81. UNEP (2011) Green economy—pathways to sustainable development and poverty eradication—a synthesis for policy makers. UNEP, Nairobi
82. United Nations (2014) Water and energy—information brief. UN. https://www.un.org/waterforlifedecade/pdf/01_2014_water_and_energy.pdf. Accessed 1 Sept 2020
83. United Nations (2015) Transforming our world: the agenda for sustainable development. United Nations, New York
84. United Nations (UN Water) (2020) Water, food and energy. United Nations. https://www.unwater.org/water-facts/water-food-and-energy/. Accessed 30 Sept 2020
85. Victor P (2008) Managing without growth: slower by design, not disaster. Edward Elgar, Cheltenham
86. Victor PA (2012) Growth, degrowth and climate change: a scenario analysis. Ecol Econ 84:206–212. https://doi.org/10.1016/j.ecolecon.2011.04.013
87. WMO (2019) The state of greenhouse gases in the atmosphere based on global observations through 2018. In: WMO greenhouse gas bulletin 2019
88. Wallace-Wells D (2019) The uninhabitable earth: life after warming. Tim Duggan Books
89. Weitzman ML (2009) On modeling and interpreting the economics of catastrophic climate change. Rev Econ Stat 91:1–19
90. Wied-Nebbeling S (2007) Grundlagen der Mikroökonomik (Springer-Lehrbuch). Springer, München
91. Winans K, Kendall A, Deng H (2017) The history and current applications of the circular economy concept. Renew Sustain Energy Rev 68:825–833. https://doi.org/10.1016/j.rser.2016.09.123
92. Witze A (2020) The Arctic is burning like never before—and that's bad news for climate change. Nature 585:336–337. https://doi.org/10.1038/d41586-020-02568-y
93. World Bank (2020) Global economic prospects. World Bank, Washington
94. World Commission on Environment and Development (WCED) (1987) Our common future. Oxford University Press, Oxford, New York
95. Xu C, Kohler TA, Lenton TM, Svenning J-C, Scheffer M (2020) Future of the human climate niche. Proc Natl Acad Sci 201910114. https://doi.org/10.1073/pnas.1910114117

The Evolution of the Water–Energy–Food Nexus as a Transformative Approach for Sustainable Development in South Africa

Stanley Liphadzi, Sylvester Mpandeli, Tafadzwanashe Mabhaudhi, Dhesigen Naidoo, and Luxon Nhamo

Abstract Water scarcity, as one of the top three risks to livelihoods, continues to dominate the global development agenda. Research is faced with the task of developing water knowledge and innovation solutions to support policy and decision-making on formulating coherent strategies that drive sustainable socio-economic development. In South Africa, the Water Research Commission (WRC) and its partners have taken the lead in providing transformative solutions for socio-economic development in the face of multiple and interlinked challenges. Although the WRC has championed integrated approaches such as Integrated Water Resources Management (IWRM), since 2011, and in line with emerging global trends, the focus has shifted to the more polycentric water–energy–food (WEF) nexus. The WRC has since instituted a WEF Nexus Lighthouse, a cross-cutting research, development, and innovation flagship programme that cuts across all its Key Strategic Areas (KSAs). Several research projects across the KSAs are linked to the WEF Nexus Lighthouse, following a defined research trajectory to inform resource management, decision-making, and policy. This study details the evolution of the WEF nexus as a polycentric and transformative framework that addresses various societal and environmental challenges in an integrated manner. We highlight the achievements made towards transforming the WEF nexus from being only a conceptual and discourse framework into an analytical decision support tool. Policy decisions that are based on WEF nexus have the potential to improve livelihoods and enhance sustainable socio-economic development for the achievement of the 2030 Global Agenda on Sustainable Development. The approach addresses complex socio-ecological by providing a strategic evidence base for informing policy- and decision-making.

S. Liphadzi · S. Mpandeli · D. Naidoo · L. Nhamo (✉)
Water Research Commission, Lynwood Manor, Pretoria 0181, South Africa
e-mail: luxonn@wrc.org.za

T. Mabhaudhi · L. Nhamo
Centre for Transformative Agricultural and Foo Systems, School of Agricultural, Earth and Environmental Sciences, University of KwaZulu-Natal, Pietermaritzburg, South Africa

S. Mpandeli
School of Environmental Sciences, University of Venda, Thohoyandou, South Africa

Keywords Adaptation · Resilience · Sustainable development · Nexus planning · Water scarcity · Transformative approaches

1   Introduction

The water–energy–food (WEF) nexus has gained prominence in recent years mainly due to its capability to integrate the management and governance of the three inter-linked sectors of food, energy, and water [60, 65]. It is a step beyond sectoral approaches to resource planning, transitioning towards a more holistic and inte-grated resource management [13]. Despite its envisaged importance as a decision support tool in natural resources management, its adoption and operationalisation into a policy framework had been hindered by lack of empirical models capable of assessing resources holistically, recognising all sectors as being influenced by every other sector [38]. However, recent developments have shown progress in transitioning the WEF nexus from a conceptual space, and facilitating discourse, into an analytical decision support tool [56]. The transition to using the WEF nexus as an analytical tool opens the doors to knowledge synthesis for informing policy- and decision-making, as well as making it operational. The recently developed WEF nexus integrative model addresses pertinent issues around sustainability that include:

(a)   understanding the complex interdependence between components within a system in space and time, focusing on the efficiency of the whole system and not only the productivity of individual sectors,

(b)   recognising the essence of cross-sectoral planning and utilisation of resources in promoting a balanced and inclusive dialogue and decision-making processes,

(c)   formulating cross-sectoral policy solutions that promote mutually benefi-cial responses that enhance cooperation among sectors, and public–private partnership at multiple scales, and

(d)   providing a model for out and up-scaling interventions [56]. These are envis-aged to contribute to the sustainability and eventual security of water, energy, and food resources, through evidence-based policy objectives that allow a sustainable and healthy ecosystem [13].

Recorded published work on the three-way mutual interlinkages among the WEF sectors only started in 2008 [29] and has since grown into a topical subject in sustain-ability circles. The approach further gained prominence at the Bonn 2011 World Economic Forum where it was introduced as an approach for achieving and managing sustainable development, and ensuring resource security [31]. To date, nexus plan-ning, in general, has become a mantra in sustainability circles as the approach has been recognised as an important tool that unpacks the intricate interdependencies among interlinked sectors [44, 56]. It has grown to be an integral part of initiatives that aim to achieve several of the Sustainable Development Goals (SDGs), particularly SDG 1 (poverty eradication), SDG 2 (zero hunger), SDG 6 (provision of water and sani-tation), and SDG 7 (access to affordable and reliable energy) (Fig. 1). The linkages with SDGs have transformed the WEF nexus into an important tool for assessing

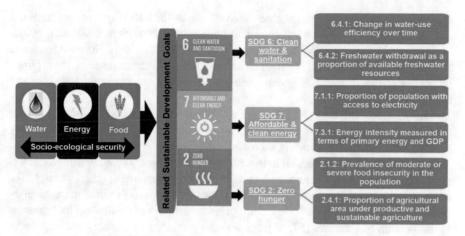

Fig. 1 Relating the WEF nexus and SDGs

progress towards the 2030 Global Agenda on Sustainable Development [56]. The linkages between the WEF nexus and the related SDGs facilitate the evaluation of country performance in SDGs implementation. Thus, nexus planning has become an important decision support tool to advance the cause of the SDGs. It has become a pathway to reduce poverty and achieve sustainable socio-ecological outcomes [10]. The essence of both the WEF nexus and the SDGs includes the following common principles:

a.   the promotion and guidance of sustainable and efficient use of resources,
b.   the access and equal distribution of resources to all, and
c.   the sustainable conservation of the natural resource base.

These WEF attributes have transformed the WEF nexus into an indispensable tool to assess SDGs implementation over time, and at the same time establishing quantitative relationships among complex but interlinked sectors [56]. The main advantage of nexus planning in resource management is its capability to assess the implementation of SDGs and to analyse trade-offs and synergies between indicators [46]. This is relevant in solving today's challenges that are also interlinked, multilateral, and multidisciplinary, requiring polycentric and transformative approaches to provide informed strategies. Monocentric approaches have the tendency of exacerbating existing challenges [44].

In South Africa, the Water Research Commission (WRC) has been spearheading the WEF nexus research through its WEF Nexus Lighthouse, a flagship research, development, and innovation programme aimed at promoting integrated planning, utilisation, and management of resources for sustainable development [39]. To date, over 30 projects across its Key Strategic Areas (KSAs) have been funded by the WRC, and results have been providing local solutions, but with a global impact. Of note, the WRC-funded research, through its partners, has defined WEF nexus indicators, developed a WEF nexus analytical model, and has applied the theory of change as a process to operationalise the WEF nexus [56]. Research results have not only benefited South Africa but have had huge global impacts as evidenced by the high number of citations of its WEF nexus-related publications. South Africa and southern Africa are now acknowledged as well ahead of other countries and regions, respectively, in terms of WEF nexus research, development, operationalisation, and implementation [43, 59].

The importance of the WEF nexus is earnestly being recognised in tandem with the evidence from research results, but real implementation has been slow, particularly because of the current prevailing sector-based governance and institutional arrangements where the three sectors sit in three different departments [39]. Current sectoral governance structures where each has its own policy documents that normally take opposite priority programmes to meet their target unintentionally threaten the security of essential resources due to conflicting priority programmes, which usually results in duplication of activities [51, 61]. Each sector pursues its mandates and policies at the detriment of the National Development Plan (NDP) Vision 2030 and SDGs [56]. However, it is imperative to note that the approach does not intend to create a mega "Nexus Ministry", but rather build a strong stakeholder platform for the coordination and facilitation of policy dialogues that would translate into the identification of priority areas for intervention. With evidence from science, such dialogues allow stakeholders to prioritise optimal projects that give the overall trade-off solutions for all sectors involved.

The need for a systematic, holistic, and integrated approach in resource management in South Africa is influenced by serious challenges of water, energy, and food insecurity. Fifty percent of the South African population does not have enough food as the country is food insecure at the household level, ninety-eight percent of the country's water supply is already allocated, and the country currently faces instability in the energy sector [86]. The distribution of natural resources in South Africa and the location of economic development hubs are uneven, a situation that further complicates management constraints and exacerbates inequality in the access to the resources [51]. The emphasis on the security of water, energy, and food is based on the fact that the three resources form the basis of a resilient economy [48]. Apart from the water scarcity challenges, South Africa also faces limited arable land (yet 20% of South Africans are food insecure) [22], highly depends on condemned coal-fired power (86% of the country's electricity is generated from coal, posing serious health risks), and depends on oil imports [26]. Estimates show that by 2030, 65% of South Africa's electricity will still be generated from coal and the population would have increased to more than 60 million, and as early as 2025 the country will face a water deficit of 1.7% [86]. At the same time, the National Development Plan (NDP) targets an increase of more than 50% of land under irrigation [69]. Indications are that the country's resources will be stretched to the limit in the near future, and this calls for urgent integrated solutions that promote resource use efficiency.

On the other hand, climate change is exacerbating the existing challenges of resource insecurity [18]. Thus, risk and vulnerability will only worsen if resources continue to be managed in sectors [56, 86]. Thus, ensuring resource use efficiency through transformative approaches is a priority for South Africa and southern

Africa, centred on improving livelihoods, building resilience, employment creation, and economic development [51, 60]. Given the importance of nexus planning in addressing complex interlinked challenges, this study aims to:

(a) synthesise the evolution of the WEF nexus in South Africa,
(b) illustrate the most significant progress made in implementing and operational-
     ising the approach,
(c) emphasise the role of nexus planning in achieving sustainability in resource
     management, and
(d) provide a practical example from research results.

Thus, this study details the evolution of the WEF nexus as a transformative frame-work for addressing distinct but interlinked sectors in an integrated manner. The premise is to provide policy- and decision-making with the relevant data to use to formulate strategies that improve livelihoods and enhance progress towards SDGs.

1.1 Defining Sustainable Development

To understand the role of the WEF nexus in assessing progress towards SDGs, it is imperative to have a basic understanding of the concepts of sustainable development and sustainability indicators. The term "sustainable development" is interchangeably used in many ways and could mean a variety of things depending on the context [35]. As a result, sustainable development has been defined in many different ways. From an ecological perspective, sustainable development is defined as the fundamental and organising base for sustaining limited resources that provide for the require-ments of future generations [49]. It is a concept meant to achieve the best from socio-ecological interactions, but at the same time ensuring a healthy environment, which on its part will continue providing adequate natural supplies to humankind [35, 49]. This definition gives emphasis on an overarching principle of sustainable development, which is the need to balance the use and protection of resources, but without compromising human life. In its report on sustainable development (the most cited document and the principal sustainable development document), the Brundt-land Commission defined sustainable development as *"development that meets the needs of the present without compromising the ability of future generations to meet their own needs"* [87]. The Brundtland definition consists of two main aspects: (1) needs (meeting human demands now and in future) and (2) limits (limitations to natural resource supply and the ability of the biosphere to meet human demands for natural resources now and in future).

In essence, the Brundtland Report emphasises that during the course of human development, the interventions and changes to the environment were initially at a small scale with limited impact, but today these interventions are threatening livelihoods [73].

Sustainable use of water-related ecosystems

SGD 6.6 targets protecting and restoring water-related ecosystems, including mountains, forests, wetlands, rivers, aquifers, and lakes. The indicator forms part of Goal 6 in that water and sanitation services are its main objective; thus while Target 6.6 is about ecosystems, it is the services of these ecosystems to society that are important. It is these ecosystems that need to be protected and managed if sustainability is to be a possibility. Ideally, the data for Target 6.6 would include the spatial extent of water-related ecosystems, the volumes of water contained by them, the quality of water within them, and the health of the ecosystems that they contain.

In short, sustainable development refers to development that does not endanger the systems that support life on earth [49]. However, this has been impossible as human interventions on the environment have altered the same environment that sustain life, triggering a host of challenges like climate change and novel socio-ecological interactions that are risking human health [5]. This is evidenced by the emergence of novel infectious diseases such as Ebola, HIV/AIDS, and COVID-19 that have claimed millions of lives [66].

Sustainable energy resources

Sustainable energy which incorporates energy access, efficiency, and renewable energy is a critical enabler for all SDGs. Globally, there has been huge progress made towards the achievement of SDG 7 as well as a shift towards sustainable energy supply options. In developing countries, there has been continued progress in access to electricity with a global increase from 83% in 2010 to 87% in 2017. Energy efficiency and use of renewable energy increased globally from 8.6% in 2010 to 10.2% in 2016. However, more effort is still required to improve access to clean and safe cooking fuels and technologies for 3 billion people globally as this has increased at a very low rate of 0.5% per annual from 2010 to 2017 [81]. In sub-Saharan Africa, there is need to increase electrification as only 44% of the population has access to electricity [80].

On the other hand, the UN Report titled "The Future We Want", which forms the basis of the SDGs, defines sustainable development as *"promoting sustained, inclusive and equitable economic growth, creating greater opportunities for all, reducing inequalities, raising basic standards of living; fostering equitable social development and inclusion; and promoting integrated and sustainable management of natural resources and ecosystems that support inter alia economic, social and human development while facilitating ecosystem conservation, regeneration and restoration and resilience in the face of new and emerging challenges"* [82]. This UN definition is

Fig. 2 Concept of
sustainable development
illustrated

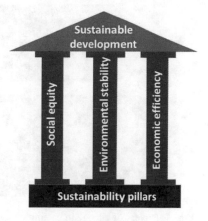

also illustrated in Fig. 2 and is the one adopted for this study. Achieving sustainable development is dependent on the balance of three pillars, mainly social equity, environmental stability, and economic efficiency (Fig. 2).

1.2 The WEF Nexus: What It Is?

The essence of the WEF nexus is based on integration, that is, seeing different aspects together [43]. In other words, it refers to the management of different but interlinked sectors in a holistic and integrated manner, and not in "silos". For example, it is recognised that food security is dependant on water and energy security, meaning that any disturbance on any one of the three sectors affects the other two [42, 59, 63]. The foundation of the WEF nexus concept is built on ensuring the security of resources, and that any developments in one sector should not compromise the security of the other sectors [56]. Thus, it is about cross-sectoral management of resources and to ensure a balanced socio-ecological system. As a transformative approach, the WEF nexus improves resource use efficiency and is a catalyst for sustainable development [56]. The capability of the WEF nexus to integrate different but interconnected sectors forms the basis of nexus planning and concerns about the security of resources. Figure 3 illustrates how the WEF nexus simplifies socio-ecological interactions and how it benefits both environmental and livelihoods security.

Sustainable food resources

The SDGs, in particular Goal 2 (*End hunger, achieve food security and improved nutrition and promote sustainable agriculture*), have included monitoring of the sustainability of food production [82]. Goal 2 includes indicators that consider the nourishment of people, food insecurity, production of food, income and sustainability of food producers and production, security of

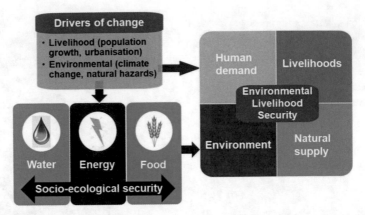

Fig. 3 Drivers of change impacting on the socio-ecological systems and how they are related to the WEF nexus

food plant and animal genomes, the orientation of governments towards food security, and finally the proper operation of food markets [82]. These are the indicators expected to monitor the overall status of food security in terms of sustainability.

SDG Target 2.3 is described as "*by 2030 double the agricultural productivity and the incomes of small-scale food producers, particularly women, indigenous peoples, family farmers, pastoralists and fishers, including through secure and equal access to land, other productive resources and inputs, knowledge, financial services, markets and opportunities for value addition and non-farm employment*". The target is measured by an indicator of productivity. SDG Target 2.4 is a direct measure of sustainable agriculture and is described as "*by 2030 ensure sustainable food production systems and implement resilient agricultural practices that increase productivity and production, that help maintain ecosystems, that strengthen capacity for adaptation to climate change, extreme weather, drought, flooding and other disasters, and that progressively improve land and soil quality*".

In recent years, nexus planning has become and has been used as an essential decision support tool for unpacking and addressing the complex and interrelated essential resources. The three WEF resources (water, energy, and food) sustain the social, economic, and environmental sustainability components (Fig. 3). In other words, it presents a framework to comprehend and systematically analyse the socio-ecological interactions and to provide coherent strategies for a more coordinated cross-sectoral management of resources. This forms the core of the concept as it guides the management of trade-offs and synergies in an integrated manner [56].

As already alluded to, the WEF nexus is a polycentric and transformative approach that ensures the security of the essential resources of water, energy, and food [53]. Therefore, the WEF nexus is a three-dimensional framework that is used either as an analytical tool, a conceptual framework, or a discourse framework [56]. Analytically, the WEF nexus is a tool that systematically applies both qualitative and quantitative methods to comprehend the intricate relationships among WEF resources; conceptually, it outlines and unpacks the complexity in the interlinkages of the WEF sectors and enhances sustainable development; as a discourse, it is used as a tool for problem-solving to enhance cross-sectoral interactions [56]. Nexus planning is also important for monitoring the performance of the WEF nexus components, as well as the SDGs [75]. These fundamental aspects of the WEF nexus have facilitated in transforming the approach into an important framework in decision-making, particularly in sustainability circles [36].

2 The Approach to Assess WEF Nexus Development

An impact study was conducted through a mixed research method, basing results from data collected and archived by the WRC, as well as research outputs mainly through publications and expert assessments. A qualitative assessment was done through a systematic review of data on innovations related to WEF nexus that was retrieved from the commission's database where it is stored and verified as part of the performance of the organisation. Innovations developed through the WRC research projects are reported by project managers and in most cases discussed and confirmed by the project reference group. Each project reference group is composed of field experts. The diverse of the reference group enables a thorough evaluation of the research results and products. Once a project is finalised, innovations are compiled and disseminated to stakeholders for uptake through stakeholder workshops and the WRC's biannual symposium. The reports are freely available from the WRC website.

The expert assessment was applied to the evolution and development of the WEF nexus in South Africa, focusing on the innovations reported by research managers and submitted to the WRC database. We focused mainly on two years' (2017/18 and 208/19) records, as it was a period that witnessed a surge in WEF nexus innovations as it evolved from a theoretical framework to a practical decision support tool and analytical model capable of enhancing synergies and minimising trade-offs between the intricate linkages of WEF resources. A quantitative assessment was done through a meta-analysis of published papers and reports originating from the WRC, highlighting the most outstanding innovations. While most of the data used are available electronically, our desktop study reviewed research reports and journal publications relating to the projects associated with the recorded innovations. The WEF nexus is a cross-cutting subject and thus is associated with work done in most KSAs and Lighthouses. The innovation report describes the innovation created, the title and contract number of the project, the field of application, journal paper, as well as the stakeholders/beneficiaries of the reported innovation. The identified WEF nexus-related innovations were compiled in an excel spreadsheet, recording information such as innovation title, objective, application area, beneficiaries, and key strategic research area. Keywords used to pick WEF nexus projects and innovations include nexus, WEF, operationalisation, and cross-sectoral.

The contribution of each innovation towards achieving SDGs was also noted and recorded (Fig. 4). This was achieved by assessing the content of each innovation report in relation to addressing the SDGs. The role of each innovation in achieving the SDGs was an important part of this study as we intended to assess and relate the contribution of the WEF nexus in assessing progress and country performance towards the SDGs. The search noted six outstanding SDGs that were being considered to be closely linked to the scope of research and innovation projects. The most mentioned SDGs in innovation reports include: 2 (zero hunger), 3 (good health and well-being), 6 (clean water and sanitation), 7 (affordable and clean energy), 11 (sustainable cities and communities), 13 (climate action).

The recorded excel data compiled were then grouped according to specific KSA (broader), as well as according to the specialised field of research (narrow). Data on innovation were rearranged according to its contribution to the WEF nexus (Fig. 4). The contribution of each innovation towards achieving Sustainable Development Goals (SDGs) was also recorded (Fig. 4). This was carried out by assessing the content of each innovation report in relation to the aspiration of the SDGs. Five SDGs, which include 1 (no poverty), 2 (zero hunger), 3 (good health and well-being), 6 (clean water and sanitation), and 7 (affordable and clean energy), were considered to be closely linked to the scope of research projects related to the WEF nexus.

The judgement as to whether an innovation belongs to a KSA or a specific specialised area of research was based on the information recorded in the project database. Table 1 was used to categorise and evaluate the impact of innovations from each project. However, the verdict regarding the contribution of each innovation to the WEF nexus and the selected SDGs was based on the scoring as indicated in

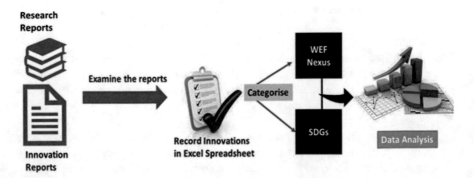

Fig. 4 Innovation data recording and handing framework used in this study

Table 1 Evaluation criteria for WEF nexus-related projects and innovations funded by the WRC

Project name/no		Start year	End year
Summary			
Project selection criterion	Directly linked to the WEF nexus	Considers elements of the WEF nexus	WEF nexus innovation
Innovation type	Conceptual framework	Discourse tool	WE nexus operationalisation
Criterion	Scope	Question	Ranking (0–5)
1. Effectiveness	Interrogates whether an intervention results in the achievement of a valued outcome of action measured in terms of units of products or services or their monetary value	Have valued outcomes been achieved?	0 = unable to evaluate 1 = least effective 5 = most effective
2. Efficiency	The amount of effort used to achieve an effectiveness determined by calculating the cost unit of product or service	How much effort was required to achieve a valued outcome?	0 = unable to evaluate 1 = least 5 = most
5. Responsiveness	The extent the innovation satisfies the needs, preferences, or values of particular groups	Do project policy outcomes transform a particular group?	0 = unable to evaluate 1 = least 5 = most
6. Appropriateness	The value or worth of a programme/project's objectives and the tenability of assumptions underlying the objectives	Are desired outcomes (objectives) actually worthy or valuable?	0 = unable to evaluate 1 = not appropriate 5 = appropriate
7. Impact	The degree to which intervention has affected communities beyond the implementation area	How many households/individuals benefitted?	0 = unable to evaluate 1 = no impact 5 = high impact
Total score			

Table 1. The final decision in this regard considered the best fit, in which the innovation is likely to make a huge contribution. When an innovation is considered to have an impact, a number one was assigned under the WEF nexus element(s) and under the specific SDG(s). Zero was assigned under elements of WEF and SDGs, in which the innovation is not making a recognisable contribution. Data were arranged and analysed using Microsoft Excel and SPSS programmes.

3   Evolution of WEF Nexus Research in South Africa

Evidence on WEF nexus research in South Africa shows that the World Wide Fund for Nature-South Africa (WWF-SA) started working on the approach in 2014 when they published a series of reports on the WEF nexus published in the same year. Popular titles of WWF-SA publications include: (a) climate change, the food–energy–water nexus and food security in South Africa [16], (b) the food–energy–water nexus: understanding South Africa's most urgent sustainability challenge [86], and (c) water, energy and food: a review of integrated planning in South Africa [27]. Although the WRC created a flagship programme in 2011 (WEF Nexus Lighthouse) to direct WEF nexus research in the country and southern Africa, it started producing positive results and making an impact after 2014. This is because of project cycles that can take up to four years.

Since then, the WRC has developed flagship innovations on the WEF nexus that include: (a) WEF nexus research needs in South Africa [39]; (b) linking the WEF nexus to resource security [40]; (c) WEF nexus and climate risks [60]; (d) WEF nexus and climate change adaptation [51]; (e) prospects for improving irrigated agriculture through the WEF nexus [41]; (f) WEF nexus as a tool to transform rural livelihoods [44]; and the development of a WEF nexus integrative model [56]. The list does not include a book that is currently in development. The most outstanding WEF nexus innovation was the development of an analytical model that establishes numerical relationships among WEF sectors and identifies priority areas for intervention [56]. This model has already been used as a decision support tool to transform rural livelihoods [44]. Another ongoing project is on operationalising the WEF nexus through the theory of change.

As already alluded to, since 2011, the WRC has been spearheading nexus research in South Africa and the southern African region with evidence of published research reports and journal articles. The WEF Nexus Lighthouse oversees WEF nexus research, development, and innovation, and the commission has premiered transformative research that includes nexus planning, circular economy, and sustainable food systems. These practical examples of transformative approaches are suitable for providing transdisciplinary solutions to closely linked but different challenges and for guiding policy- and decision-making on coherent climate change adaptation strategies.

3.1   Overall WEF Nexus Innovations in Each KSA

The total number of innovations produced by research projects falling under the three KSAs in the WRC is presented in Fig. 5. During the 2017/18 financial year, thirty-four innovations were recorded, of which nineteen were associated with the

Fig. 5 Water innovations developed under the three WRC Key Strategic Areas (KSAs) between 2017 and 2019

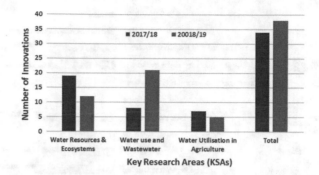

Water Resources and Ecosystems KSA, nine were from the Water Use and Wastewater Management, and seven were developed by the Water Use in Agriculture KSA (Fig. 5).

In 2018/19, another thirty-four innovations have so far been recorded, of which twelve are from the Water Resources and Ecosystems, twenty-one are from the Water Use and Wastewater Management, and five are from the Water Use in Agriculture (Fig. 5). More innovations were created or produced in 2018/19 than in the previous year (2017/18), which is attributed to increased production of innovations in Water Use and Wastewater KSA. This is significant as more innovations are expected by the end of 2019. The most outstanding innovations include (a) the development of an integrative WEF nexus model, (b) computing of water–energy–food nexus index, (c) evaluating the state of the WEF nexus in South Africa, (d) framing desalination within the water–energy–climate nexus, (e) a framework to operationalise the WEF nexus through the theory of change, (f) a framework for improving rural livelihoods through the WEF nexus, (g) developing guidelines for WEF nexus implementation in southern Africa, and (h) COVID-19 and beyond, building resilience through nexus planning and transformative approaches.

The fact that more innovations are now being produced by various KSAs is based on the WRC's focus, which now emphasises more on project demonstrations. This forms part of the Fourth Industrial Revolution theme championed by the South African government. The WRC is complementing this theme through its research and development agenda.

Figure 6 shows a graphical presentation of innovations developed under various Specialised Research Areas (SRAs). SRAs are programmes or thrusts under the key strategic research areas. While most of the innovations were developed in the Aquatic Ecosystems Health, and Biodiversity SRAs in 2017/18, a remarkable number of innovations were created in specialised research areas on Water Quality and Economic Uses of water (all with >4 innovations).

In 2018/19, more innovations were created under the sanitation specialised research area than in the previous year. The sanitation research area produced more innovations in 2018/19 than any other SRA (Fig. 6). No innovation was reported for new specialised research areas (Water Governance and Water Uses and Community Initiatives). Innovations are a result of research collaboration among teams who work

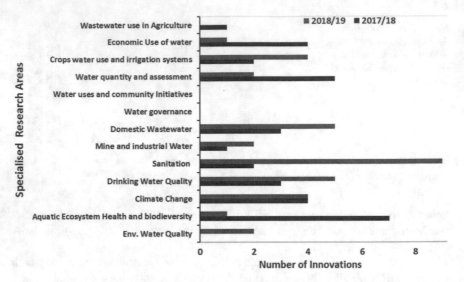

Fig. 6 Innovations produced by various key specialised research areas

together for between 3 and 5 years, but in this case, the two programmes are still new and under development, which explains why there were no innovations reported from these two programmes. The impact and importance of each innovation were based on the criteria shown in Table 1.

3.2 Milestones and Research Gaps in WEF Nexus Research

The WRC-led research on WEF nexus shows a clear trajectory, where the initial work was mainly on the literature review, presenting the approach as a conceptual framework and as a discourse tool. However, current research has shifted towards transitioning the approach into an analytical tool, as well as an operational framework. Initial WEF nexus research funded by the WRC did well in presenting the approach as a conceptual framework that simplifies human understanding of the complex linkages among WEF resources and promoting coherence in policy-making, enhancing sustainable development. As a discourse tool, the approach has been well portrayed as a problem-framing concept that promotes cross-sectoral collaboration [42, 51, 56, 60]. This initial research managed to unpack the intricacies of the WEF nexus as a decision support tool for integrated resources management and poverty alleviation, vulnerability reduction, promotion of inclusive development and regional integration, as well as building resilient communities in the advent of climate change [51, 60].

Several research outputs in the form of publications were made during this theoretical stage, but evidently lacking on the analytics component to transform the WEF nexus into a decision support tool that provides real-world solutions and demonstrates what the approach says it can do. Thus, the gap that remained to be filled was the development of WEF nexus evaluation tools and models capable of assessing synergies and trade-offs in an integrated manner, and capable of eliminating "silos" in resource management, and averting conflicts, minimising investment risks, and maximising on economic returns [32, 38]. The absence of empirical evidence attracted its criticism due to lack of clarity and practical applicability [15]. Some critics had gone further by branding the approach as a repackaging of the IWRM [7]. Some research went further to depict the approach as an ambitious and ambiguous programme that will never achieve its intended goals [89].

Buoyed by the challenges of transforming the WEF nexus into a practical and evidence-based operational framework, the WRC has since moved into supporting research that aims to develop WEF nexus analytical models and pathways to operationalise the approach from its present state of being mainly theory. This was motivated by national needs such as ensuring simultaneous water, energy, and food security, job, and wealth creation, and expands the irrigated land in an era of water and energy scarcity [69]. The WRC, through its partners, has since defined WEF nexus sustainability indicators and developed an integrative model to assess composite indices to facilitate WEF resources performance, monitoring, and evaluation through a multi-criteria decision-making (MCDM) process [56]. As the model is based on indicators, the WEF nexus has thus become an important tool for monitoring the performance of SDGs, particularly SDGs 2 (zero hunger), 6 (clean water and sanitation), and 7 (affordable and clean energy), as also suggested by other studies [75]. The WRC has driven the WEF nexus research agenda and has so far developed a South African-based WEF nexus framework, developed considering pertinent issues facing the country (Fig. 7) [39]. Figure 7 indicates that an acceptable state of human livelihood, well-being, and environmental sustainability is achievable when there are right policies, strategies, as well as alternative clean, renewable options that are in place.

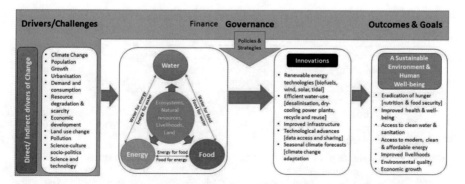

Fig. 7 WEF nexus framework for South Africa to attain simultaneous water, energy, and food security with particular emphasis on Sustainable Development Goals 2, 6, and 7

An important feature of the WEF nexus is its capability to assess the performance of SDGs and act as a decision support tool to achieve sustainability in resource planning, utilisation, and development, particularly SDGs 2, 6, and 7. The nexus framework given in Fig. 4 explains and clarifies the connectedness of the three WEF sectors. The direct and indirect drivers of change, affecting these intricate linkages, are also demonstrated in Fig. 7. The essence of the WEF nexus consists of the drivers that are critical to the WEF sectors, and the cross-sector feedbacks in South Africa, the reason why they are placed in the centre of the nexus. The framework forms the research agenda for the WEF nexus in South Africa. The innovations expected from the research agenda are in direct response to the WRC Knowledge Tree components like knowledge and innovation, capacity building through the application of the integrative analytical model, and the formulation of integrative policy and legal frameworks to drive the political will. The whole process would culminate in improved livelihoods, resilient communities, environmental sustainability, and inclusive development.

3.3   The WEF Nexus Integrative Model

Table 2 provides the selected WEF nexus sustainability indicators and pillars as defined by Nhamo et al. [56]. The MCDM was used to integrate and determine WEF indicators and compute indices through the analytic hierarchy process (AHP), a MCDM method [71, 78]. The AHP is a method for measuring ratio scales from distinct and continuous paired comparisons [70]. This is essential for decision-makers to formulate priorities and deduce the best possible decisions. Indicators and pillars (Table 2) are essential for determining quantitative relationships among sectors by means of a comparison matrix, which provides a value to each indicator. This is achieved by comparing and relating each indicator to other indicators, assigned an

Table 2 Sustainability indicators and pillars for WEF nexus sectors

Sector	Indicator	Pillars
1. Water	Proportion of available freshwater resources per capita (availability) Proportion of crops produced per unit of water used (productivity)	Affordability Stability Safety
2. Energy	Proportion of the population with access to electricity (accessibility) Energy intensity measured in terms of primary energy and GDP (productivity)	Reliability Sufficient Energy type
3. Food	Prevalence of moderate or severe food insecurity in the population (self-sufficiency) Proportion of sustainable agricultural production per unit area (cereal productivity)	Accessibility Availability Affordability Stability

Source Nhamo et al. [56]

index value according to Saaty's AHP pairwise comparison matrix (PCM), and then normalised to have the indicator [70, 71].

The AHP was used to numerically relate indicators through the pairwise comparison matrix (PCM) (Eq. 1). This theory (AHP) has a comparison matrix that compares two factors/indicators at a time, applying a scale ratio of between 1/9 and 9 (Table 3) [71]. An array between 1 and 9 represents a significant relationship, and a range between 1/3 and 1/9 symbolises a less important relationship. These ranges are well illustrated and described in Table 3.

A ranking of 9 shows that in relation to the column factor, the row factor is 9 times more important. On the contrary, a rating of 1/9 indicates that relative to the column indicator, the row indicator is 1/9 less important. But in instances where the column and row indicators are equally important, they have a rating of 1. In WEF nexus, the indices are dependent on the impact of the indicator on its overall scoring.

The PCM was used to determine indices, taking the eigenvector (a vector with the same direction even if a linear transformation is applied) conforming to the largest eigenvalue (the rank of the eigenvector) of the pattern (matrix), and thereafter normalising the total of the factors or indices [76]. Figure 8 shows a graphical presentation of the processes followed when applying the AHP.

Table 3 Fundamental scale for pairwise comparisons

Intensity of importance	Definition	Explanation
1	Equal importance	Elements *a* and *b* contribute equally to the objective
3	Moderate/weak importance of one over another	Experience and judgement slightly favour element *a* over *b*
5	Essential or strong importance	Experience and judgement strongly favour element *a* over *b*
7	Demonstrated importance	Element *a* is favoured very strongly over *b*; its dominance is demonstrated in practice
9	Absolute importance	The evidence favouring element *a* over *b* is of the highest possible order of affirmation
2, 4, 6, 8, 1/2, 1/4, 1/6, 1/8	Intermediate values between the two adjacent judgements	When compromise is needed. For example, 4 can be used for the intermediate value between 3 and 5
1/3	Moderately less important	
1/5	Strongly less important	
1/7	Very strongly less important	
1/9	Extremely less important	
Reciprocals of above nonzero	If *a* has one of the above numbers assigned to it when compared with *b*, then *b* has the reciprocal value when compared with *a*	

Source Saaty [71]

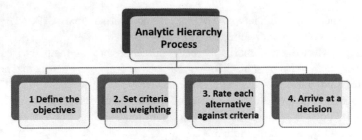

Fig. 8 Procedure for the analytic hierarchy process

The PCM method is important for estimating the priority weight for each indicator as related to several decision criteria and alternatives. The priority weight vector is on the right triangle of the PCM w [72]. The overall value or ranking of each indicator is then estimated. The basic input is the pairwise pattern or matrix, A, of n criteria, developed based on Saaty's scaling ratios (Table 3), which is of the order $(n \times n)$ [64]. A is a pattern with elements a_{ij}. The pattern generally is reciprocal, and it is mathematically expressed as:

$$a_{ij} = \frac{1}{a_{ij}} \tag{1}$$

After deriving this pattern, it is then normalised as pattern B, in which B is the normalised pattern of A, with elements b_{ij} and mathematically expressed as:

$$b_{ij} = \frac{a_{ij}}{\sum_{j=1} n a_{ij}} \tag{2}$$

Each weight value w_i is calculated as:

$$w_i = \frac{\sum_{j=1} n b_{ij}}{\sum_{i=1} n \sum_{j=1} n b_{ij}}, \quad i, j = 1, 2, 3 \ldots, n \tag{3}$$

The composite WEF nexus index is then computed as a weighted average of all the estimated indices. This integrated composite index indicates the overall resource management and progress towards sustainability in resource use in an integrated manner. The composite WEF nexus index establishes the numerical relationship of different but interlinked sectors of water, energy, and food, whose performance is interpreted using Table 4, which categorises the indices into either sustainable or unsustainable.

The model has already been successfully applied as a tool to assess transforming rural livelihoods in southern Africa [44]. The study applied the WEF nexus analytical livelihood model with complex systems understanding to assess rural livelihoods, health, and well-being in southern Africa. The indices were used to develop a spider graph showing the present performance category in resource management in the

Table 4 WEF nexus indicator performance classification categories

Indicator	Unsustainable	Marginally sustainable	Moderately sustainable	Highly sustainable
Water availability (m³/per capita)	<1700	1700–6000	6001–15,000	>15,000
Water productivity (US$/m³)	<10	10–20	21–100	>100
Food self-sufficiency (% of pop)	>30	15–29	5–14	<5
Cereal productivity (kg/ha)	<500	501–2000	2001–4000	>4000
Energy accessibility (% of pop)	<20	21–50	51–89	90–100
Energy productivity (MJ/GDP)	>9	6–9	3–5	<3
WEF nexus composite index	0–09	0.1–0.2	0.3–0.6	0.7–1

Source Nhamo et al. [56]

region. The spider graph indicates a clear imbalance, uneven, and unsustainable resource management (Fig. 9). There is an extra emphasis on food security (food self-sufficiency) at the detriment of the other sectors. Sector-based approaches are the main reason for the irregular shape of the spider graph, as there are limited opportunities to mitigate trade-offs. The good performance in food self-sufficiency (an index of 0.357) also comes with the detrimental effect of allocating more water resources towards agriculture (which stands at 62%), a situation that is affecting other competing sectors. There is clear evidence of more resources being directed towards the agricultural sector, yet water and hydropower remain in short supply. Thus, Fig. 9 identifies priority areas for interventions [56].

Fig. 9 Present performance in resource management in southern Africa. The irregular spider graph (the orange centrepiece) indicates whether resource management is sustainable or unsustainable

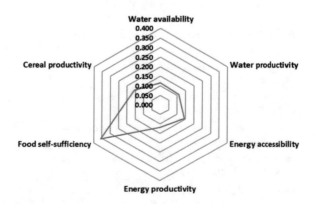

The informed evidence from the WEF nexus integrative model provides opportunities to operationalise the WEF nexus through the theory of change, an inclusive stakeholder engagement meant to reach an informed common goal [88].

3.4   Overview of WEF Nexus Indicators (2015 and 2018)

The WEF nexus analytical model has simplified the interpretation and understanding of the complex interactions between the interlinked WEF resources. This quantitative evidence of the interconnectedness of the three resources provides the basis to identify priority sectors for intervention that would lead to sustainability [57]. An outline of the statuses of WEF nexus-related indicators in South Africa in 2015 and 2018 is given in Table 5 [90]. These statuses, together with expert advice, as well as the pillars are considered during the pairwise comparison [56]. The AHP, a multi-criteria decision-making (MCDM) process, was also used to normalise the PCM values using Eqs. (2) and (3), and to establish the composite indices for each indicator and the integrated WEF nexus index. The PCM for each of the reference years was used to normalise indices, and calculate the CR. The procedure is preferred as it is successfully used in other fields such as economics, sociology, programming, sustainable development, environmental studies, land suitability mapping, town planning, strategic planning, resource allocation, and project and risk management to integrate distinct factors and monitor progress towards specific targets [20, 48, 56].

The estimated indices (Table 6) are presented as a spider diagram (Fig. 10), which gives a clear illustration and synopsis of the changes that occurred in resources use and management between 2015 and 2018. This capability of the WEF nexus

Table 5 Overview of the WEF nexus indicators for South Africa

Indicator	Indicator status		
	2015	2018	Units
Proportion of available freshwater resources per capita (availability)	821.3	821.4	m^3
Proportion of crops/energy produced per unit of water used (water productivity)	26.2	26.2	$/m^3$
Proportion of population with access to electricity (accessibility)	85.5	84.4	%
Energy intensity measured in terms of primary energy and GDP (productivity)	8.7	8.7	MJ/GDP
Prevalence of moderate/severe food insecurity in the population (self-sufficiency)	5.7	6.2%	%
Proportion of sustainable agricultural production per unit area (cereal productivity)	3.5	5.6	kg/ha

Source World Bank Indicators [90]

Table 6 WEF nexus composite indices for South Africa

Indicator	Composite indices	
	2015	2018
Water availability	0.126	0.099
Water productivity	0.128	0.221
Energy accessibility	0.141	0.079
Energy productivity	0.111	0.199
Food self-sufficiency	0.314	0.292
Cereal productivity	0.180	0.111
WEF integrated index	0.155	0.203

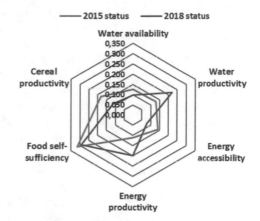

Fig. 10 An outline in the progress towards SDGs between 2015 and 2018 in South Africa. The irregular shape of the spider graphs indicates an unsustainable economy

analytical model to assess progress in resources management over time facilitates the assessment of progress towards SDGs.

The numerical outline shown in the spider diagram highlights the interactions, interconnectedness, and interdependencies of the intricate relations between the WEF sectors as seen together, providing an outlook and general progress in SDGs implementation. The more an indicator is away from the centre of the axis, the better the sustainability level of that indicator, or the closer it is to the nucleus the worst the sustainability level. The irregular shapes of the spider diagrams (the centrepieces) for both years are an indication that South Africa still must do more to achieve sustainability in resource use and management. The irregular shape of the graphs represents disparities in resource planning, allocation, utilisation, and management, which mostly favour the rich. Such a scenario usually exacerbates existing challenges, promotes inequality, aggravates poverty, and triggers resource insecurity. These imbalances are normally triggered by sectoral management of resources, as each sector pursues its own strategies.

For both reference years, South Africa was concentrating or putting much emphasis on food security (food self-sufficiency). In 2018, the country had a better water productivity index; however, this progress was attained at the expense of the other indicators. This is evidenced by the contraction experienced in the other indicators, a situation that gave the irregular shape of the graphs (Fig. 10). The resulting phenomenon is that when food security and water productivity were improving, the other indicators were contracting (Fig. 10), an indication of sectoral resource management and reactive measures in case of disasters, which often result in unsustainability.

Without compromising the food security indicators, as well as the progress achieved in water productivity, the country has to allocate resources to those indicators that seem to be stable as instability will manifest in the short term [56]. One indicator of a sustainable and balanced economy is when the centrepiece or the spider graph becomes circular. On the other hand, when the shape of the spider graph is irregular it indicates an unsustainable economy and unequal distribution of resources. In a WEF nexus environment, all sectors are considered equal and should be developed in an integrated manner.

The better performance in water productivity in 2018 is understood to have resulted from the measures that were taken during and after the drought that occurred in the 2015/16 rainy season [54]. The same applies to energy productivity that improved in the same period, possibly as a result of the measures implemented during the energy crisis of 2017. However, the same sectoral measures resulted in low energy accessibility as compared to 2015 (decreased by 1.3% between 2015 and 2018). Therefore, "the good performance" in food self-sufficiency and water and energy productivity mainly resulted from the measures that were undertaken after a crisis, but the same measures just transferred challenges to the other sectors which later performed poorly during the same period. Without informed WEF nexus interventions, achieving sustainability will prove difficult. Reactive and uniformed interventions normally compromise adaptation and resilience-building initiatives.

As already alluded to, sustainable resource management requires that all indicators attain the highest index of the "best" performing indicator, without compromising other indicators, resulting in a circular shape of the spider graph. By contrast, the current sectoral approach will continue to create an imbalance in the economy and retard development, consequently slowing progress towards achieving the 2030 agenda [55]. Balanced resource management may suggest that resources are being managed holistically to achieve sustainability but can still be classified as unsustainable if the indices remain at low values. An assessment of the changes in SDGs implementation taking place over time provides evidence to research and decision-makers on integrating strategies aimed at operationalising the WEF nexus in order to manage resources holistically from a nexus perspective. Cross-sectoral approaches in resource management usually translate to cost savings compared to the replication of developmental projects by competing sectors [50].

The spider graph discloses the strengths and weaknesses of a country and at the same pointing the priority areas that need interventions. These capabilities have made the WEF nexus a valuable adaptation and decision support tool for assessing progress towards national and international goals like the NDP and SDGs. Interventional scenarios are then developed to inform policy- and decision-making on pathways to achieve sustainability. Thus, the WEF nexus has been recognised as an important approach for tracking the utilisation and management of resources at a given time. The WEF nexus has become an important polycentric decision support tool that unpacks and frames complex socio-ecological interactions. However, further research is needed on developing scenarios to inform decision-making on balancing resource management and achieve the 2030 Global Agenda on Sustainable Development.

3.5 Interpretation of the Indices

An outline of a country's performance in the utilisation and development of resources, together with the dynamics and transformations over time, is presented as a spider diagram (Fig. 10). The centrepiece highlights the strengths of a country and identifies priority areas that need intervention. The procedure allowed to quantitatively interpret the interlinkages among related sectors, through the AHP whose indices range between 0 and 1. An index of 0 represents unsustainable resource management and 1 represents the best sustainability. The basis of the WEF nexus analytical model is to assess, monitor, and track resource utilisation and performance, and it acts as a guide in developing scenarios to achieve the desired sustainability. These are the attributes that link the WEF nexus with SDGs. Depending on the composite index, a country can set targets to achieve the desired balance in resource management to ensure stability and security towards sustainable development. Although a country can be ranked according to its WEF nexus integrated index, the aim is to move towards sustainable development and attain the 2030 Global Agenda of the SDGs in line with national strategies and plans, rather than a comparison with other countries.

The results presented on the spider graph provide decision-makers with evidence and indicate where interventions are critically needed. Realistic targets are then set as guided by the analytical framework. In this case study of South Africa, interventions could target the provision of safe and reliable water, and clean and safe energy and improve on crop productivity. This is particularly relevant for South Africa as it states in the National Development Plan (NDP) to increase the area under irrigation by 149,000 ha in order to ensure food security [69]. Nexus planning provides a platform for stakeholder engagement, which facilitates the formulation of coherent strategies that guide policy- and decision-making. This is of fundamental importance as the process identifies synergies and trade-offs at the right time among for prompt interventions.

3.6 Applications of the WEF Nexus Analytical Framework

The WEF nexus has evolved into a multi-purpose and multi-centric framework used as either a conceptual framework, discourse tool, analytical framework or an assessment tool [12, 33, 77]. The development of WEF nexus analytical framework has paved the way for a host of opportunities that include, among others:

assessing the performance and progress of SDGs, policy framing, strategies for climate change adaptation, livelihood transformation, project appraisal, governance structures, among other applications. The approach assesses the effects of different intervention scenarios in the development and management of resources for informed planning by decision-makers. Applications of the WEF nexus analytical framework are diverse and multi-scale, and they include:

1. **Scenario planning**: The impacts of future climate change are unpredictable, and climate change projections are embedded with uncertainty, a situation that complicates policy- and decision-making in the context of climate change [37]. As a result, decision-makers prefer using scenario planning in policy formulation for climate change, as it uses alternatives and response options in climate change adaptation [17, 74]. As scenario planning uses the "what if, how and when" criteria rather than a predictive approach which has lots of uncertainty, the WEF nexus analytical framework comes in handy as it analyses and assesses all possible intervention strategies and indicates whether intervention on any resource is sustainable or not, and how the intervention affects the other resources [9, 60]. As climate change is a cross-cutting challenge, adaptation strategies and scenario planning should be formulated around the WEF nexus framework to ensure that policies are evidence-based, well coordinated, targeted, and coherent. Both the WEF nexus and scenario planning are recognised for their ability to not relying on or assigning likelihoods to future conditions; instead, they widen decision priorities for informed potential responses [74]. Linking the WEF nexus and scenario planning provides pathways towards flexible and reliable opportunities to upscale or downscale as informed by prevailing challenges. Scenario planning methods include the Shared Socioeconomic Pathways (SSPs), the Representative Concentration Pathways (RCPs), and Representative Agricultural Pathways (RAPs). Scenario planning facilitates an integrated analysis of both long-term and near-term modelling analyses of events to understand and assess the degree of vulnerability. The assessment allows to recommend adaptation and mitigation strategies [67, 83, 84]. Linking scenario planning to the WEF nexus allows the reflexivity, resilience, responsiveness, and revitalisation of adaption policies. Reflexivity refers to the ability to continuously deal with a host of challenges as they come; resilience refers to the capability to address and adapt to new environments in the midst of uncertainties; on the other hand, responsiveness refers to the capability to address persistent demands and expectations; revitalisation is the reignition of policies and ensures continuous application [14].

2. **Rural livelihood transformation**: Demand and competition for resources have been increasing exponentially due to population growth and migration, economic growth, international trade, urbanisation, diversifying diets, cultural and technological changes, as well as climate variability and change [19]. These societal megatrends are exerting pressure on already depleting natural resources, threatening their sustainability going forward. The challenges have widened the gap between rural and urban communities, particularly in developing countries

where rural communities continue lacking basic services and amenities such as safe water and sanitation, clean energy, and balanced and nutritious diets [21, 30]. Disparities in resources distribution between rural and urban communities, as well as the insecurities of water, energy, and food, require a cross-sectoral approach in resource management, development, and utilisation like the water–energy–food (WEF) nexus, which facilitates inclusive and equitable development as well as coordinated resource planning and management [60]. The implementation of the WEF nexus analytical framework creates synergies and ensures that trade-offs are avoided, and at the same time eliminating the duplication of activities, which is often common in a sectoral approach to resource development [60]. The current sectoral approach only exacerbates the vulnerabilities of mostly rural people who lack resources to adapt to change [28]. The transformation of rural livelihoods and the sustainability of adaptation strategies are hinged on the understanding of the role of the WEF nexus in framing effective policies and institutions to ensure sustainable livelihood transformation. Adopting the nexus approach at the regional level enhances transboundary cooperation strategies, regional integration, and poverty alleviation [51].

3.   **Assessing performance and progress of Sustainable Development Goals (SDGs)**: As progress on SDGs is assessed through quantifiable indicators that keep track of the measurable targets, the WEF nexus analytical framework becomes a "fitting approach" for monitoring and evaluating the progress towards the 2030 Agenda of the SDGs. The WEF nexus analytical framework assesses, monitors, and tracks resource utilisation and performance over time using the same SDGs indicators. As a cross-sectoral approach, the WEF nexus integrates indicators across sectors and elucidates how best resources can be allocated between competing needs, thus making the implementation of SDGs more efficient and cost-effective. The WEF nexus approach is, thus, an approach for understanding the complex and dynamic interlinkages between the issues related to the securities of water, energy, and food, making it directly linked to SDGs 2, 6, and 7, as well as for guiding the interactions between the natural environment and the biosphere [10, 25, 51, 60].

4.   **Project appraisal**: The WEF nexus analytical framework is a systematic approach that integrates water, energy, and food in project planning and appraisal at all levels and is essential in preventing trade-offs, as well as helping to enhance the development potential. Thus, the WEF nexus is essential in project design and evaluation, especially for complex projects that involve various expertises and are transboundary or transnational in nature. The WEF nexus analytical framework project leaders and funders to quantify the linkages among WEF nexus components identify critical links and leverage the results to improve project design and implementation [77]. An analysis of nexus interactions at the planning and design stage assists funders to understand the long-term impact of investments across different sectors and users [85]. The approach permits not only the identification of possible trade-offs, such as the risk of the unsustainability of some crucial project aspects but also synergies, whereby investment in

one sector benefits the others. Thus, the nexus approach ensures that investments in the project are responsible, achievable, and sustainable.

4 Transitioning Towards Sustainable Food Systems Through Nexus Planning

A sustainable food system refers to an agriculture system that delivers healthy food to meet current food requirements, while at the same time preserving healthy and sustainable ecosystems that are capable of providing food for generations to come, with controlled negative impact to the environment [2]. It is a system that encourages local production and knowledge, providing for nutritious and healthy food, which is available, accessible, and affordable to all and at all times, and at the same time protecting farmers and workers, consumers, and communities [23]. Thus, a sustainable food system is a complex system driven by intricately interlinked economic, social, cultural, and environmental factors, which require transformative thinking and integrated assessment tools to guide informed strategic polices [2]. Nexus planning provides integrated tools to establish relationships between complex systems, indicating areas for priority intervention [45]. The approach is a lens for understanding the complex interlinkages among economic, social, and ecological interactions and is a transitional pathway towards sustainable food system [52].

Nexus planning balances the interactions between agriculture, water, and energy resources, making it particularly relevant for southern Africa where there is an evident imbalance in the use of resources [45, 56]. Currently, 95% of the agricultural land in the region is rainfed, yet there is perennial food insecurity [8, 58]. This calls for investment in agriculture development to tap into the potential of the region to be food self-sufficient, but this requires considerations for land and water management solutions to ensure sustainability. Nexus planning provides decision support pathways towards ecological sustainability on irrigation [68]. Apart from investment and technological developments in the agriculture sector, nexus planning improves productivity and resource use efficiency [34]. For example, the development of smart plants has seen the emergence of more drought-tolerant varieties through genetic modification and genome editing [79]. Some plants are also being engineered to develop more efficient photosynthetic pathways that fully use the sun's available energy [62]. These technologies are envisaged to improve productivity in hot climates and water-scarce regions. Remote sensing has equally become an important component of irrigation management, particularly for irrigation scheduling [1]. Remote sensing products are being used to pinpoint areas of wet and dry zones in cultivated fields, and for estimating crop water requirements [3]. On the other hand, mobile apps and other social media platforms are being used to disseminate information on weather, rainfall, and soil humidity to allow better farm management and productivity, as well as information on markets [4]. These advances need to be applied in a way that is detrimental to the environment.

The rationale is based on establishing a quantitative relationship among the intricately connected and multiple interactions within the agricultural systems. This facilitates an understanding of how socio-economic, environmental, and ecological interactions are influencing change and unsustainability. This is fundamental for evaluating the major societal outcomes that are affected by these interactions (food security, ecosystem services, and social welfare) [24]. Figure 11 presents a comprehensive conceptual framework illustrating processes of an agriculture value chain, including the role of the nexus planning in transitioning towards sustainable food systems. Nexus analytical modelling is a preferred approach in integrated analyses, using sustainability indicators to provide quantitative relationships among intricately connected components of an agricultural system beyond food production, processing, distribution, and consumption, which also include economic, social, and environmental drivers (Fig. 11).

As food systems are complex social-ecological systems that involve various interactions between human (economic and political trends, food price volatility, population dynamics, changes in diets and nutrition, and advances in science and technology) and natural (landcover changes, land and soil degradation, climate change, biodiversity loss, sea-level rise, and air pollution) components [24, 47], it is paramount to understand these relationships holistically to transition towards sustainable food systems (Fig. 11). In between, the social-ecological systems are external drivers, which include exposure and sensitivity, that also drive food systems. Knowledge of these drivers and how they relate to influence activities and outcomes of food systems is important for informing policy decisions [6]. Food and nutrition security are the major outcomes of any food system; thus, a food system is considered vulnerable or resilient when it fails to deliver food security or ensures food security,

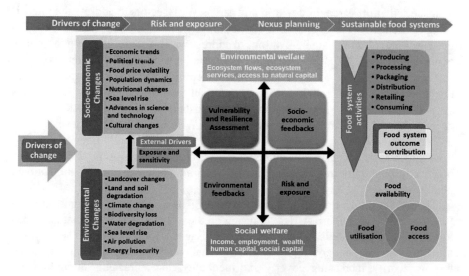

Fig. 11 A sustainable food system conceptual framework illustrating the processes and interactions involved in achieving a sustainable food system through nexus planning

respectively [24]. Nexus planning connects these interactions by defining, measuring, and modelling progress towards sustainability, through a set of indicators formulated around resource utilisation, accessibility, and availability [56]. These developments facilitate modelling, monitoring, and simulating some aspects of sustainability.

The conceptual framework (Fig. 11) is concerned with the creation of a food system that uses resources efficiently and reduces food waste at every stage from primary production to transformation and consumption. Nexus modelling develops knowledge-based tools to assess vulnerability and resilience, as well as recovery options and the potential of a food system. The tools facilitate the identification of pathways for simultaneous food security and resource conservation through an analysis of the food system activities and outcomes, integrating socio-ecological, political, and economic factors summarised in socio-economic and global environmental change drivers (Fig. 11). This is based on the understanding that food systems are socio-ecological systems comprising biophysical and social factors that are linked through feedback mechanisms [11, 24, 47].

Identifying and modelling the fundamental and complex processes of a food system through nexus planning ensures that food and nutritional outcomes are preserved or enhanced over time and across generations. This is achieved by identifying priority areas for intervention in a timely manner, allowing decision- and policy-makers to track progress in resources management and implement policies that enhance positive transformations.

5 Recommendations and Conclusions

A concerted effort is required within and across WEF sectors to address the intensifying challenges of resource security in South Africa and southern Africa. There is a growing body of promising innovations to address water, energy, and food insecurity generated by the WRC. The WRC has highlighted several times that there is a slow movement of the inter-connectivity between sector interventions and trade-offs for resource management, and this needs to change if South Africa is to achieve targets set in the National Development Plan and the Sustainable Development Goals. The development of a research agenda for informing WEF nexus research, development, and innovation forms a key part of the WRC's strategy to fund targeted interventions aimed at (i) increasing the evidence base, (ii) demonstrating the WEF nexus in practice through pilot studies, and (iii) mainstreaming research outputs into national policy.

This synthesis has traced the evolution of WEF nexus research and innovation in South Africa and southern Africa, focusing on the results of the work that is funded by the WRC. The work is motivated by the envisaged importance of the WEF nexus in addressing national and regional priority challenges such as water, energy, and food insecurity, increasing poverty and inequality, and increasing unemployment and urbanisation. The WRC-funded WEF nexus research has produced some interesting innovations that are driving the nexus from a theoretical framework into a practical and operational decision support tool that provides evidence and basis for sustainable resource management strategies based on integrated quantification of water, energy, and food resources simultaneously. Of note is the development of a WEF nexus integrative and analytical model that simplifies the understanding of the complex linkages among the WEF resources, vividly showing how resources are managed and indicating priority areas for intervention. The analytical model has, thus, managed to quantify interlinkages for the three interlinked resources, reduce trade-offs and enhance synergies for optimum resource use and development, enable an assessment of the sustainability of water, energy, and food systems, and apply different scenarios across ecological and socio-economic zones and spatial scales. These innovations have managed to provide evidence and drive policy changes that facilitate the transitioning of the WEF nexus into a practical and operational framework essential to build resilience for sustainable development. More research is needed on developing context-based scenarios that eliminate uncertainty brought about the present sector-based approaches in resource management.

References

1. Adeyemi O, Grove I, Peets S, Norton T (2017) Advanced monitoring and management systems for improving sustainability in precision irrigation. Sustainability 9(3):353
2. Allen T, Prosperi P (2016) Modeling sustainable food systems. Environ Manage 57(5):956–975
3. Alvino A, Marino S (2017) Remote sensing for irrigation of horticultural crops. Horticulturae 3(2):40
4. Amarnath G, Simons GWH, Alahacoon N, Smakhtin V, Sharma B, Gismalla Y, Mohammed Y, Andriessen MCM (2018) Using smart ICT to provide weather and water information to smallholders in Africa: the case of the Gash River Basin, Sudan. Clim Risk Manag 22:52–66
5. Bellard C, Bertelsmeier C, Leadley P, Thuiller W, Courchamp F (2012) Impacts of climate change on the future of biodiversity. Ecol Lett 15(4):365–377
6. Béné C, Oosterveer P, Lamotte L, Brouwer ID, de Haan S, Prager SD, Talsma EF, Khoury CK (2019) When food systems meet sustainability—current narratives and implications for actions. World Dev 113:116–130
7. Benson D, Gain AK, Rouillard JJ (2015) Water governance in a comparative perspective: from IWRM to a 'nexus' approach? Water Altern 8(1)
8. Besada H, Werner K (2015) An assessment of the effects of Africa's water crisis on food security and management. Int J Water Resour Dev 31(1):120–133
9. Bieber N, Ker JH, Wang X, Triantafyllidis C, van Dam KH, Koppelaar RHEM, Shah N (2018) Sustainable planning of the energy-water-food nexus using decision-making tools. Energy Policy 113:584–607
10. Biggs EM, Bruce E, Boruff B, Duncan JMA, Horsley J, Pauli N, McNeill K, Neef A, Van Ogtrop F, Curnow J (2015) Sustainable development and the water-energy-food nexus: a perspective on livelihoods. Environ Sci Policy 54:389–397
11. Binder CR, Hinkel J, Bots PWG, Pahl-Wostl C (2013) Comparison of frameworks for analyzing social-ecological systems. Ecol Soc 18(4)
12. Bizikova L, Roy D, Swanson D, Venema HD, McCandless M (2013) The water-energy-food security nexus: towards a practical planning and decision-support framework for landscape investment and risk management. International Institute for Sustainable Development (IISD), Winnipeg

13. Boas I, Biermann F, Kanie N (2016) Cross-sectoral strategies in global sustainability governance: towards a nexus approach. Int Environ Agreem Polit Law Econ 16(3):449–464
14. Breeman G, Dijkman J, Termeer C (2015) Enhancing food security through a multi-stakeholder process: the global agenda for sustainable livestock. Food Secur 7(2):425–435
15. Cairns R, Krzywoszynska A (2016) Anatomy of a buzzword: the emergence of 'the water-energy-food nexus' in UK natural resource debates. Environ Sci Policy 64:164–170
16. Carter S, Gulati M (2014) Climate change, the good energy-water nexus and food security in South Africa. In: Understanding the food-energy-water nexus. World Wide Fund for Nature-South Africa (WWF-SA), Cape Town
17. Cobb AN, Thompson JL (2012) Climate change scenario planning: a model for the integration of science and management in environmental decision-making. Environ Model Softw 38:296–305
18. Conway D, Schipper ELF (2011) Adaptation to climate change in Africa: challenges and opportunities identified from Ethiopia. Glob Environ Change 21(1):227–237
19. Corvalan C, Hales S, McMichael AJ, Butler C, McMichael A (2005) Ecosystems and human well-being: health synthesis. World Health Organization
20. Dizdaroglu D (2017) The role of indicator-based sustainability assessment in policy and the decision-making process: a review and outlook. Sustainability 9(6):1018
21. Dos Santos S, Adams EA, Neville G, Wada Y, De Sherbinin A, Mullin Bernhardt E, Adamo SB (2017) Urban growth and water access in sub-Saharan Africa: progress, challenges, and emerging research directions. Sci Total Environ 607:497–508
22. du Toit DC, Ramonyai MD, Lubbe PA, Ntushelo V (2011) Food security. Department of Agriculture, Forestry ad Fisheries (DAFF), Directorate Economic Services, Pretoria
23. Eakin H, Connors JP, Wharton C, Bertmann F, Xiong A, Stoltzfus J (2017) Identifying attributes of food system sustainability: emerging themes and consensus. Agric Hum Values 34(3):757–773
24. Ericksen PJ (2008) Conceptualizing food systems for global environmental change research. Glob Environ Change 18(1):234–245
25. FAO (2014) The water-energy-food nexus: a new approach in support of food security and sustainable agriculture. Food and Agriculture Organisation of the United Nations (FAO), Rome
26. Freeman J (2015) A critical look at the moral case for fossil fuels. Energy LJ 36:327
27. Goga S, Pegram G (2014) Water, energy and food: a review of integrated planning in South Africa. In: Understanding the food-energy-water nexus. World Wide Fund for Nature-South Africa (WWF-SA), Cape Town
28. Harvey CA, Rakotobe ZL, Rao NS, Dave R, Razafimahatratra H, Rabarijohn RH, Rajaofara H, MacKinnon JL (2014) Extreme vulnerability of smallholder farmers to agricultural risks and climate change in Madagascar. Phil Trans R Soc B 369(1639):20130089
29. Hellegers P, Zilberman D, Steduto P, McCornick P (2008) Interactions between water, energy, food and environment: evolving perspectives and policy issues. Water Policy 10(S1):1–10
30. Hemson D, Meyer M, Maphunye K (2004) Rural development: the provision of basic infrastructure services. Human Sciences Research Council (HSRC), Pretoria
31. Hoff H (2011) Understanding the nexus. Background paper for the Bonn 2011 conference: the water, energy and food security nexus. Stockholm Environment Institute, Stockholm
32. Howells M, Hermann S, Welsch M, Bazilian M, Segerström R, Alfstad T, Gielen D, Rogner H, Fischer G, Van Velthuizen H (2013) Integrated analysis of climate change, land-use, energy and water strategies. Nat Clim Change 3(7):621
33. Keskinen M, Guillaume JHA, Kattelus M, Porkka M, Räsänen TA, Varis O (2016) The water-energy-food nexus and the transboundary context: insights from large Asian rivers. Water 8(5):193
34. Klerkx L, Rose D (2020) Dealing with the game-changing technologies of agriculture 4.0: how do we manage diversity and responsibility in food system transition pathways? Glob Food Secur 24:100347
35. Kuhlman T, Farrington J (2010) What is sustainability? Sustainability 2(11):3436–3448

36. Leck H, Conway D, Bradshaw M, Rees J (2015) Tracing the water–energy–food nexus: description, theory and practice. Geogr Compass 9(8):445–460
37. Lemos MC, Rood RB (2010) Climate projections and their impact on policy and practice. Wiley Interdiscip Rev Clim Change 1(5):670–682
38. Liu J, Yang H, Cudennec C, Gain AK, Hoff H, Lawford R, Qi J, de Strasser L, Yillia PT, Zheng C (2017) Challenges in operationalizing the water–energy–food nexus. Hydrol Sci J 62(11):1714–1720
39. Mabhaudhi T, Simpson G, Badenhorst J, Mohammed M, Motongera T, Senzanje A, Jewitt A, Naidoo D, Mpandeli S (2018) Assessing the state of the water-energy-food (WEF) nexus in South Africa. Water Research Commission (WRC), Pretoria
40. Mabhaudhi T, Chibarabada T, Modi A (2016) Water-food-nutrition-health nexus: linking water to improving food, nutrition and health in sub-Saharan Africa. Int J Environ Res Public Health 13(1):107
41. Mabhaudhi T, Mpandeli S, Nhamo L, Chimonyo VGP, Nhemachena C, Senzanje A, Naidoo D, Modi AT (2018) Prospects for improving irrigated agriculture in southern Africa: linking water, energy and food. Water
42. Mabhaudhi T, Mpandeli S, Chimonyo VGP, Nhamo L, Backeberg G, Modi AT (2016) Propsects for improving irrigated agriculture in southern Africa: linking water, energy and food. In: 2nd world irrigation forum (WIF), Chiang Mai, 6–8 Nov 2016
43. Mabhaudhi T, Mpandeli S, Luxon Nhamo VGP, Chimonyo AS, Naidoo D, Liphadzi S, Modi AT (2020) Emerging water-energy-food nexus lessons, experiences, and opportunities in southern Africa. In: Vasel-Be-Hagh A, Ting DSK (eds) Environmental management of air, water, agriculture, and energy. CRC Press, Florida, p 141
44. Mabhaudhi T, Nhamo L, Mpandeli S, Naidoo D, Nhemachena C, Sobratee N, Slotow R, Liphadzi S, Modi AT (2019) The water-energy-food nexus as a tool to transform rural livelihoods and wellbeing in southern Africa. Int J Environ Res Public Health 16(2970)
45. Mabhaudhi T, Nhamo L, Mpandeli S, Nhemachena C, Senzanje A, Sobratee N, Chivenge PP, Slotow R, Naidoo D, Liphadzi S (2019) The water-energy-food nexus as a tool to transform rural livelihoods and well-being in southern Africa. Int J Environ Res Public Health 16(16):2970
46. Mainali B, Luukkanen J, Silveira S, Kaivo-Oja J (2018) Evaluating synergies and trade-offs among sustainable development goals (SDGs): explorative analyses of development paths in South Asia and sub-Saharan Africa. Sustainability 10(3):815
47. Marshall G (2015) A social-ecological systems framework for food systems research: accommodating transformation systems and their products. Int J Commons 9(2)
48. Matchaya G, Nhamo L, Nhlengethwa S, Nhemachena C (2019) An overview of water markets in southern Africa: an option for water management in times of scarcity. Water 11(5):1006
49. Mensah J, Ricart Casadevall S (2019) Sustainable development: meaning, history, principles, pillars, and implications for human action: literature review. Cogent Soc Sci 5(1):1653531
50. Mpandeli S, Nhamo L, Moeletsi M, Masupha T, Magidi J, Tshikolomo K, Liphadzi S, Naidoo D, Mabhaudhi T (2019) Assessing climate change and adaptive capacity at local scale using observed and remotely sensed data. Weather Clim Extremes 26:100240
51. Mpandeli S, Naidoo D, Mabhaudhi T, Nhemachena C, Nhamo L, Liphadzi S, Hlahla S, Modi A (2018) Climate change adaptation through the water-energy-food nexus in southern Africa. Int J Environ Res Public Health 15(2306)
52. Newman RJS, Capitani C, Courtney-Mustaphi C, Thorn JPR, Kariuki R, Enns C, Marchant R (2020) Integrating insights from social-ecological interactions into sustainable land-use change scenarios for small Islands in the western Indian Ocean. Sustainability 12(4):1340
53. Nhamo L, Chilonda P (2012) Climate change risk and vulnerability mapping and profiling at local level using the household economy approach (HEA). J Earth Sci Clim Change 3(123):2
54. Nhamo L, Mabhaudhi T, Modi AT (2019) Preparedness or repeated short-term relief aid? Building drought resilience through early warning in southern Africa. Water SA 45(1):75–85
55. Nhamo L, Mabhaudhi T, Mpandeli S (2019) A model to integrate and assess water-energy-food nexus performance: South Africa case study. In: World irrigation forum (WIF), Bali, 1–7 Sept 2019

56. Nhamo L, Mabhaudhi T, Mpandeli S, Dickens C, Nhemachena C, Senzanje A, Naidoo D, Liphadzi S, Modi AT (2020) An integrative analytical model for the water-energy-food nexus: South Africa case study. Environ Sci Policy 109(109):15–24. https://doi.org/10.20944/prepri nts201905.0359.v1
57. Nhamo L, Mabhaudhi T, Mpandeli S, Nhemachena C, Senzanje A, Naidoo D, Liphadz S, Modi AT (2019) Sustainability indicators and indices for the water-energy-food nexus for performance assessment: WEF nexus in practice—South Africa case study. Preprint 2019050359:17
58. Nhamo L, Matchaya G, Mabhaudhi T, Nhlengethwa S, Nhemachena C, Mpandeli S (2019) Cereal production trends under climate change: impacts and adaptation strategies in southern Africa. Agriculture 9(2):30
59. Nhamo L, Ndlela B, Nhemachena C, Mabhaudhi T, Mpandeli S, Matchaya G (2018) The water-energy-food nexus: climate risks and opportunities in southern Africa. Water 10(5):567
60. Nhamo L, Ndlela B, Nhemachena C, Mabhaudhi T, Mpandeli S, Matchaya G (2018) The water-energy-food nexus: climate risks and opportunities in southern Africa. Water 10(567):18. https://doi.org/10.3390/w10050567
61. Ntombi R-T (2017) Fragmented sectoral approach on water, energy and food in South Africa. Trade & Industrial Policy Strategies (TIPS), Johannesburg
62. Ort DR, Merchant SS, Alric J, Barkan A, Blankenship RE, Bock R, Croce R, Hanson MR, Hibberd JM, Long SP (2015) Redesigning photosynthesis to sustainably meet global food and bioenergy demand. Proc Natl Acad Sci 112(28):8529–8536
63. Pérez-Escamilla R (2017) Food security and the 2015–2030 sustainable development goals: from human to planetary health: perspectives and opinions. Curr Dev Nutr 1(7):e000513
64. Rao MSVC, Sastry SVC, Yadar PD, Kharod K, Pathan SK, Dhinwa PS, Majumdar KL, Sampat Kumar D, Patkar VN, Phatak VK (1991) A weighted index model for urban suitability assessment—a GIS approach. Bombay Metropolitan Regional Development Authority, Bombay
65. Rasul G, Sharma B (2016) The nexus approach to water–energy–food security: an option for adaptation to climate change. Clim Policy 16(6):682–702
66. Reperant LA, Osterhaus ADME (2017) AIDS, Avian flu, SARS, MERS, Ebola, Zika… what next? Vaccine 35(35):4470–4474
67. Riahi K, Van Vuuren DP, Kriegler E, Edmonds J, O'Neill BC, Fujimori S, Bauer N, Calvin K, Dellink R, Fricko O (2017) The shared socioeconomic pathways and their energy, land use, and greenhouse gas emissions implications: an overview. Glob Environ Change 42:153–168
68. Rockström J, Williams J, Daily G, Noble A, Matthews N, Gordon L, Wetterstrand H, DeClerck F, Shah M, Steduto P (2017) Sustainable intensification of agriculture for human prosperity and global sustainability. Ambio 46(1):4–17
69. RSA (2011) National development plan vision 2030: our future make it work. National Planning Commission (NPC), Pretoria
70. Saaty RW (1987) The analytic hierarchy process—what it is and how it is used. Math Model 9(3–5):161–176
71. Saaty TL (1977) A scaling method for priorities in hierarchical structures. J Math Psychol 15(3):234–281
72. Saaty TL (1990) Eigenvector and logarithmic least squares. Eur J Oper Res 48(1):156–160
73. de Sherbinin A, Carr D, Cassels S, Jiang L (2007) Population and environment. Annu Rev Environ Resour 32:345–373
74. Star J, Rowland EL, Black ME, Enquist CAF, Garfin G, Hoffman CH, Hartmann H, Jacobs KL, Moss RH, Waple AM (2016) Supporting adaptation decisions through scenario planning: enabling the effective use of multiple methods. Clim Risk Manag 13:88–94
75. Stephan RM, Mohtar RH, Daher B, Irujo AE, Astrid Hillers J, Ganter C, Karlberg L, Martin L, Nairizi S, Rodriguez DJ (2018) Water–energy–food nexus: a platform for implementing the sustainable development goals. Water Int 43(3):472–479
76. Stewart S, Thomas MOJ (2006) Process-object difficulties in linear algebra: eigenvalues and eigenvectors. In: Proceedings of the 30th conference of the international group for the psychology of mathematics education, Prague

77. Terrapon-Pfaff J, Ortiz W, Dienst C, Gröne M-C (2018) Energising the WEF nexus to enhance sustainable development at local level. J Environ Manag 223:409–416
78. Triantaphyllou E, Mann SH (1995) Using the analytic hierarchy process for decision making in engineering applications: some challenges. Int J Ind Eng Appl Pract 2(1):35–44
79. Tripathi L, Ntui VO, Tripathi JN (2019) Application of genetic modification and genome editing for developing climate-smart banana. Food Energy Secur 8(4):e00168
80. UN (2019) The sustainable development goals report 2019. Unted Nations (UN), New York
81. UNECE (2017) Global tracking framework: UNECE progress in sustainable energy. United Nations Economic Commission for Europe (UNECE), New York and Geneva
82. UNGA (2015) Transforming our world: the 2030 agenda for sustainable development. In: Resolution adopted by the General Assembly (UNGA). United Nations General Assembly, New York
83. Valdivia RO, Antle JM, Rosenzweig C, Ruane AC, Vervoort J, Ashfaq M, Hathie I, Tui S-K, Mulwa R, Nhemachena C (2015) Representative agricultural pathways and scenarios for regional integrated assessment of climate change impacts, vulnerability, and adaptation. In: Rosenzweig C, Hille D (eds) Handbook of climate change and agroecosystems: the agricultural model intercomparison and improvement project (AgMIP) integrated crop and economic assessments, part 1. Imperial College Press, London, pp 101–145
84. Van Vuuren DP, Edmonds J, Kainuma M, Riahi K, Thomson A, Hibbard K, Hurtt GC, Kram T, Krey V, Lamarque J-F (2011) The representative concentration pathways: an overview. Clim Change 109(1–2):5
85. Villamayor-Tomas S, Grundmann P, Epstein G, Evans T, Kimmich C (2015)The water-energy-food security nexus through the lenses of the value chain and the institutional analysis and development frameworks. Water Altern 8(1):735–755
86. Von Bormann T, Gulati M (2014) The food-energy-water nexus: understanding South Africa's most urgent sustainability challenge. In: South Africa: WWF-SA. World Wide Fund for Nature (WWF), Cape Town
87. WCED (1987) Our common future: report of the World Commission on Environment and Development (WCED). Oxford University Press, Oxford and New York
88. Weiss CH (1995) Nothing as practical as good theory: exploring theory-based evaluation for comprehensive community initiatives for children and families. In: Connell JP (ed) New approaches to evaluating community initiatives: concepts, methods, and contexts. Aspen Institute, Washington, pp 65–92
89. Wiegleb V, Bruns A (2018) What is driving the water-energy-food nexus? Discourses, knowledge and politics of an emerging resource governance concept. Front Environ Sci 6(128). https://doi.org/10.3389/fenvs.2018.00128
90. World-Bank Indicators (2020) World development indicators 2020. World Bank, Washington DC, USA

Integrated Watershed Management
Vis-a-Vis Water–Energy–Food Nexus

Seyed Hamidreza Sadeghi and Ehsan Sharifi Moghadam

Abstract Iran, like most of the developing countries in the world, faces many issues with limiting water, energy, and food resources. On the other hand, the state of the watersheds is being destroyed due to the imbalance between human and the environment leading to different environmental, social, economic, and political consequences. This requires a change of mindset so that we can understand what we do every day on the water–energy–food (WEF) nexus. Balance establishment among various main components of the ecosystems based on the WEF nexus approach is considered as one of the important parts of integrated and adaptive watershed management. This chapter briefly discusses on the concept, standards, and methods of the WEF nexus. An example of the application of the WEF nexus will also be presented from Iran exploring the interconnections between chapters of the nexus and integrated watershed management approaches. From the results and existing evidence, it can be concluded that the WEF nexus would be a practical, adjustable, and adaptive approach for fulfilling increasing demands of the human while there is a glance to limited resources. Based on the available literatures and the case study done by the authors, it can be ultimately concluded that the adoption of the WEF nexus would facilitate better achievements in integrated management of the watersheds.

Keywords Energy security · Food security · Integrated watershed management · Soil and water conservation · Sustainable development · Water–energy–food nexus

S. H. Sadeghi (✉) · E. Sharifi Moghadam
Department of Watershed Management Engineering, Faculty of Natural Resources and Marine Sciences, Tarbiat Modares University, 46417-76489 Noor, Iran
e-mail: sadeghi@modares.ac.ir

E. Sharifi Moghadam
e-mail: ehsan.sharifi@modares.ac.ir

E. Sharifi Moghadam
Natural Resources and Watershed Management Office, Firoozkooh Township, Tehran Province, Iran

© The Author(s), under exclusive license to Springer Nature Singapore Pte Ltd. 2021
S. S. Muthu (ed.), *The Water–Energy–Food Nexus*,
Environmental Footprints and Eco-design of Products and Processes,
https://doi.org/10.1007/978-981-16-0239-9_3

1 Introduction

Three commodities of water, energy, and food (WEF) are defined as the principles on which worldwide security, financial condition, and justice rest. There is now increasing grounds that anxiety in these areas threatens communities' security and encourages social conflicts [1]. Water, energy, and food are important resources for achieving economic improvements and social aspirations leading to sustainable global economic development. These elements are completely interrelated and they significantly depend on each other [95]. Any scheme that concentrates on a sector of the WEF nexus without reckoning its interrelationships has important unintended upshots. The interdependent relationships of these three critical resources are termed as WEF nexus. The relationship between water, energy, and food suggests that water handiness, energy generation/intake, and food security are highly correlated, and that action in each region affects one or both [33]. As McCarl et al. [54] stated, the target of focusing on the WEF nexus in totality is to recognize and capitalize on the WEF synergies and advantages of coordinated actions related to uncoordinated actions. Presently, the WEF nexus is completely prominent in the domain of environmental management. Dealing with resource issues related to WEF nexus require a systems view to connect human and natural systems for recognizing feasible solutions [45, 51]. Miscellaneous investigations have been reported on WEF nexus in different regions (e.g., [5, 18, 46, 71, 75, 77, 78]). On the other hand, since its promotion, the WEF nexus has related to hundreds of scientific publications. Sharifi Moghadam et al. [72] reviewed and analyzed the existing literatures on the WEF nexus approach at different scales and suggested supplementary ideas for better applicability of the WEF nexus framework for integrated management of the watershed. The review showed 10 articles had a close linkage with water–food, 49 with water–energy, 119 with water–energy–food, six with water–food–energy–ecosystems, five with water–energy–land–food, three with food–energy–environment, three with water–soil–waste, and eight with climate. However, a closer look at former endeavors gave away that the issue of WEF relevance has only been raised in the last decade. A systematic and comprehensive approach is therefore needed to protect critical resources (i.e., water, energy, and food). In this vein, integrated watershed management (IWM) is a consistent, incorporated, and multi-sectorial system scheme to management that targets to improve the productiveness and wholeness of ecosystems in the field of water, soil, plants, and animals in one area. Thus, such approaches protect ecosystem services to conserve and restore environmental and socioeconomic benefits [82]. This chapter found that using an integrated perspective for managing water, energy, and food resources in the watershed management is a new approach. This chapter presents a brief overview of the importance of IWM on three types of resources (i.e., water, energy, and food).

2 Concept of the Water–Energy–Food (WEF) Nexus

Various global initiatives and forums of international agencies focusing on the nexus framework, concentrated on water, energy, and food research. Many scientific gatherings were held in 2011–2012, for leading the concept of the WEF nexus, particularly during the preparation phase for Rio + 20 in June 2012. One relevant framework was developed on the Water, Energy, and Food Security Nexus: Solutions for the green economy as a part of the Bonn 2011 Nexus Conference as suggested in Fig. 1.

After holding the conferences, the nexus of specific parts or elements depending on the amplitude and focus of each variable is specified research or investigation and differs from the classic nexus of the three parts mentioned above. The target of this framework is to suggest a new nexus-oriented approach, which is required to present unstable growth models and imminent resource objections in order to improve security of access to primary services [13, 33]. In the Bonn 2011 Conference, it was accepted that the concept nexus, which reappeared as the new term in development disclosures, could appropriately define the interrelationship among three resources. Since then, this issue has become a very controversial topic in most international forums. However, these communities are more focused on water, energy or water–food or energy–food. The Bonn Conference of 2011 propelled the water–energy–food nexus approach into international discussions on sustainable development [33, 40]. Another approach to WEF was presented by the World Economic Forum in 2011 as shown in Fig. 2. The intention of this theoretical account is to provide situation for decision-makers to recognize the risks better, therefore in times of facing crises they will be capable to respond proactively and mobilize quickly [13].

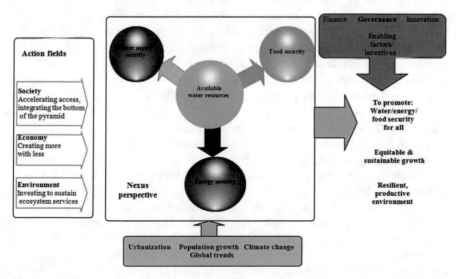

Fig. 1 Framework suggested for the Bonn 2011 Nexus Conference on the water, energy, and food security nexus (after [13, 33])

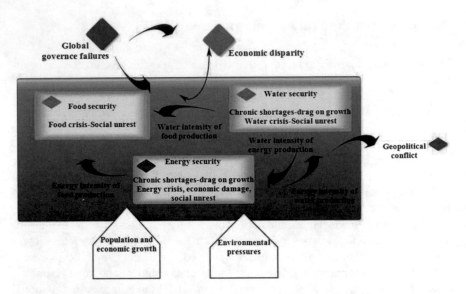

Fig. 2 Approach to WEF suggested by the World Economic Forum (after [13])

The main focus of these enterprises was to enhance the WEF nexus by promoting consciousness, highlighting the urgency challenges related to WEF, preparing international conferences, and recommending guidelines and investment suggestions [13]. Three years later, during the Sustainability Nexus water–energy–food Conference, a call was made by policy and research communities around the world to develop strategies to take a holistic approach to communication. The sustainable development goals (SDGs) also call for an integrated nexus approach. Eventually, the UN Secretary General Ban Ki-moon emphasized on applying a nexus approach, and incorporation of environmental, social and economic aspects [16]. In order to achieve the advantages of nexus approach respecting the efficiency use of the resources, it is essential to recognize, operationalize, and practice water–energy–food as the elements of nexus. International Atomic Energy Agency (IAEA) conducted a research based on integrated modeling. According to the results of this study, the authors came to this conclusion that considering water–energy–food nexus areas leads to more acceptable resources allocation, enhanced economic effectiveness, less environmental and health influences, as well as an enhanced economic progress conditions. Actually, a more broad data on relative inadequacy and productivity of the resource, and on the capacity for intensifying sustainability on various regions is provided by an integrated view across nexus [33]. In order to identify the interconnectedness types among systems, various combinations consisting of water, food, energy, trade, climate, and population growth are being surveyed. The results realizing multidimensionality and complexity of the issue were creation of those nexus [58]. Nonetheless, there is not a vast knowledge and publication in this area, and the role of water–energy–food nexus in representation of competing needs and promoting adaptation and progress is not completely recognized.

2.1 Nexus Approach

The concept of nexus is defined as a process of joining actions of different stakeholders from various elements for gaining sustainable enhancement. The term nexus is in adjustment with the framework of integrated water resource management (IWRM). In contrast to IWRM which is water sector-oriented and limit its synergy with other sector, nexus is more opened to any other sectors by encouraging capability in using resources [86]. The IWRM is a revelation in moving into a nexus framework and is useful in developing the nexus. Using IWRM, the link was built as a two-way negotiation, a neutral platform that is not water-based, but where the water community, the energy community, and the food community can discuss resource assignation as a link between resource relationships [57]. IWRM and WEF communication approaches seem similar but disagree in certain subjects. However, the supreme goals of both are promoting better use of resources to allow communities to create environmentally, socially, and economically sustainable modes [10]. The general properties of the water security nexus and IWRM have also been summarized in Table 1.

The nexus is suggested as a method of analysis for indicating the relationship among nexus sectors, viz. water, energy, and food [19]. The nexus of food, energy, and water has been depicted in Fig. 3.

Scholars and practitioners recognize nexus problems variously, and therefore, the concept is often applied as a buzzword with various meanings [15]. The core elements of nexus approach are human environmental dimensions, and they are related to ecological stability and the well-being of future generations depend on these elements. In contrast to single-factor for water, land, energy, and food, the nexus solutions should explicitly determine the decreased costs and increased benefits for both human and nature and it demands an integrated management approach [11]. As Hagemann and Kirschke [28] stated, the nexus concept does not have and

Table 1 General properties of the water security nexus and IWRM [10]

Aspect	Nexus	IWRM
Integration	Integrating water, energy, and food policy objectives	Integrating water with other policy objectives
Optimal governance	Integrated policy solutions multi-tiered institutions	Good governance principles
Scale	Multiple scales	Watershed scale
Participation	Public–private partnerships–multi-stakeholder platforms for increasing stakeholder collaboration	Stakeholder involvement in decision-making multiple actors including women
Resource use	Economically rational decision-making Cost recovery	Efficient allocations
		Cost recovery
		Equitable access
Sustainable development	Securitization of resources	Demand management

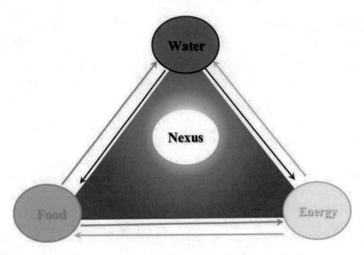

Fig. 3 Pictorial demonstration of food, energy, and water nexus (after [3])

analytical framework for integrated scientific analysis. The nexus, while providing a comprehensive understanding of the unintended aftermaths of policies, technologies, and practices, is a multidimensional tool for scientific inquiry that desires to delineate the complexity and nonlinearity of anthropogenic interactions [34]. In fact, resolutions which are based the nexus-wide applications are likely to prepare better outcomes in comparison with individual elements [54]. Applying a nexus approach in order to enhance water resources stability in energy and food supply chains is considered as a propitious approach. A nexus conceptualization can also give the transition to a green economy, which collaboratively aims to increase resource skillfulness and outstanding policy cohesiveness. Communication in the growing demand between resources (e.g., water for energy and energy for water) is another important characteristic, which refers to possibilities and synergies to increase the overall efficiency of resource use and perhaps transposition between assets. Accordingly, Hoff [33] made it clear that established policy and decision-making on the ground therefore need a way of approaching that reduces trade and creates synergies between sectors. Nexus approaches in cities include integrated framework planning for water, wastewater, and energy. Cities need to develop strategies with their inland areas and watersheds by providing agricultural products as well as reprocessing waste, e.g., by descending water uses (i.e., reusing lower-quality water for aims that need lower water quality) and by enhancing nexus approaches through peri-urban agriculture and landscaping. According to the findings of Hoff [33], the target of nexus reasoning is to magnify the mutuality of water, energy, and food security and natural sources that construct the water, soil and land security. It also tries to identify integration and determine liabilities for enhancing resource productivity, greater policy consistency and integrated natural resources management. According to Kattelus et al. [37], statements enhanced interconnectedness recognition between the sectors, lead to enhanced co-benefits across sectors and scales for further mutual trust and goals. Previous studies recommend that a new nexus-based scheme is required to deal with unsustainable growth patterns [33] and addressing causes rather than the symptoms, by determining impressive points for intervention in underlying structures and systems [87]. Just as other upcoming concepts, however, the nexus approach is still under development and passes infant stage.

2.2 WEF Nexus Components

In this chapter, the main constituents of the water–energy–food (WEF) relationship were identified. The components can be summarized as follows:

2.2.1 Water–Food

Water is a vital element for raising kinds of vegetable, fruits, cereals, animals, and also their products that we use as food and producing beverages [3]. Water for food production accounts for around 70% of water withdrawals. From another point of view, the demand for food will increase significantly as a result of the raising of population growth, urbanization and economic growth along with changes in diet [14]. Since water is the core element in the nexus, therefore investigations about water issues is the most important matter in order to reach development results in these three interconnected elements [27]. Water plays as a state thing and at the same time a control other things and is centrally linked [33]. The progress of agricultural practices globally is influenced greatly by the handiness and accessibility of water assets. Agriculture in addition to be the cause of water pollution is the victim as well. The type of crops planted, the crop cycles and the irrigation method used vary from dry to humid regions of the world. Nowadays, the relationship between water and food is a symbol of vulnerability on two foreparts. First, changing water supply patterns that affect the reliability of high-water sectors, including agriculture, and second, growing competition for limited water resources to meet the planned addition in food need [23]. Water is required for food production, especially for irrigation and crops processing. In the world, the main water consumer is agricultural, and it consumes about 90% of global freshwater in the past century [38]. Scientists estimated that 60% raise in food production may be required to reach world food need by 2050 in comparison with 2007, and therefore, it demands an increase in productivity use up to 70 million ha of additional arable land. Interim, water exploitation for agriculture, which currently consumes about 70% of global water, is supposed to increase slightly by 5% by 2050, thanks to improved productivity [52]. Reaching increasing need for water and food demand precise risk management and conditions which are related to the connection between various food and water security attributes. Several food security concerns are influenced by accessibility to water with acceptable quality. In order to produce food and further downstream effects during preparing food, it is necessary to prepare and consume sufficient water with acceptable quality. Likewise,

Table 2 Summary of risks and impacts within the water–food nexus [23]

	Risks	Impacts
Water-related risks to food security	Increased variability in water availability, particularly due to climate change Regional concentration of food production and consumption	Changes in supply of food products, leading to higher price volatility, further compounded by regional concentration of food production activities
	Impact of water quality on food production and consumption	Utilization of poor-quality water along different stages of the food supply chain can have negative impacts, including soil degradation and accumulation of contaminants within the food chain
Food-related risks to water security	Impact of agricultural activities on water resources	Use of external inputs for agriculture and food production can lead to water pollution affecting downstream activities and aquatic life
	Poorly regulated agricultural foreign direct investments (e.g., international land leasing)	Increased agricultural land leasing, when poorly regulated, could lead to expanded use of local water resources, with negative local socioeconomic impacts
	Water resource over-utilization due to food security ambitions	Pursuit of food security ambitions can strain water resources, often leading to substantial depletion in freshwater reserves

the intensification of some food production methods, for example, the more aggressive use of nutrient-enriched soils or evolving diets (e.g., the growing demand for protein-rich diets containing meat). It has significant implications for water security [23]. Nonetheless, the problem existing in the agriculture sector is fully complicated and miscellaneous each of which needs specific attentions and appropriate solutions. Table 2 shows a summary of the risks and effects of the water–food relationship.

2.2.2 Water–Energy

Guesses on the standards of progress and real need by any provided date differ mainly among different resources. According to the Global Energy Statistical Yearbook, there is waste in both global energy and electricity (e.g., in 2018, 2.3% and 5.3%, respectively). This yearbook lists China and the USA as the countries that consume the most energy and electricity in comparison with other countries in the world [4, 20]. In the last decade, the need of water becomes so important because of both population and economic growth. In 2019, it was reported that the world population is around 7.75 billion. It is forecasted that the population growth would increase to 9.7 billion in 2050 [4]. The consequence of this population growth is a high pressure on the available water and energy system, because most regions of the world have deficits in supply. Besides, the environmental junctures caused by unsustainable water and energy use are expected as the most important threat, worldwide [19]. During the period in which world is bearing environmental and resource stresses, connection between limited resources attracted the attention of researches. Determining the connection between the main resource parts and enhancing their efficiency is considered as a win-win strategy for current, and more importantly, future generation human well-being and environmental stability. Among these linkages, the water–energy nexus has been quickly risen [88]. The popularity of the concept could be traced back to the World Economic Forum in 2008, where the global challenges in terms of economic growth were identified from the water–energy nexus viewpoint [19, 84]. Water and energy are multifaceted problems with many parts changing their supply and demand. Water and energy sources are twisted and are both important for economic growth. The relationship between water and energy resources can be developed into the concept of the water–energy nexus (WEN) [22]. WEN describes the undeniable linkage between water and energy, which is the basis of clever community substructure [4, 49]. Scientist claim that informed water and energy planning and recognition of potential choices for both policy and technology can be supported better by the WEN analysis. Healy et al. [31] stated that it is expected this process can help policy makers and resource managers for conservation and stability of water and energy. Energy and water are interconnected because these elements are main consumers of one another. Water plays an important role in almost every stage of energy development, consisting of the exploitation, production and processing of fossil fuels, electricity generation, extraction, treatment and processing of waste from energy-related measures. At the same time, energy is needed to lift, move, distribute, and purify water [63, 95]. Energy requirements in transporting water have been summarized in Table 3.

Furthermore, there are six interrelated aspects to water and energy management, as shown in Fig. 4.

In addition to the innate interrelation between water and energy, it is also important to pay attention to the negative impact of energy sources on water quality. Both surface and ground water resources are under the effect of the potential of water contamination caused by energy resources including tailing seepages, fracturing fluids and

Table 3 Energy requirements (kWh) in delivering unit volume of clean water (m^3) [58]

Lake or river	0.37
Groundwater	0.48
Wastewater treatment	0.62–0.87
Wastewater reuse	1.00–2.50
Seawater	2.58–8.50

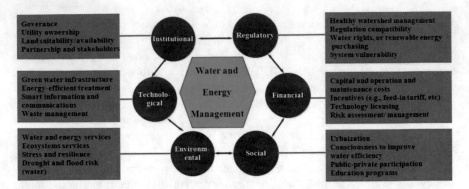

Fig. 4 Key components of water and energy management based on six interrelated aspects (after [62])

flow-back of oil and gas, emptying and seepage of acid mine in coal mining areas [49]. However as Wijayanti et al. [86] stated, the interrelation between water and energy parts can be shown in concept, but the feedback links are sophisticated and influenced by external factors. Likewise, a great deal of attention has recently been drawn to various aspects of the energy–water relationship and its importance has been gradually recognized. The water–energy nexus studies focus on energy generation from crops evaluated by Water Footprint and Life Cycle Assessment (LCA) [61], sustainable development [93], regional development and integration by water and energy security [49], methods and tools for macro-assessment [17], and attempts and possibilities for ameliorating water efficiency in the cooling systems of ther-moelectric power plants [62]. Such focuses are being paid to quantify trade-offs of the nexus, an initiative according to water and energy flow and optimization of water–energy nexus in reservoir systems [44], to demonstrate the application and interpretation of the temporal evolution to detect significant trends and regime shifts [8], to identify and understand the different water and energy consumption patterns of cities [21, 83], to classify and estimate values of water-related energy consump-tion [88], interactions across two types of virtual resource transfer networks [89], for optimal usage of the energy consumption in water–energy systems at a community scale [56], and understand the complex connection between water and energy to develop strategies since the time of sustainable water management [86].

Although the issue of water–energy observed the focus of research attention in recent years, yet there is big knowledge gaps, and limited researches about water–energy are available. In fact, there is still a limited understanding of the interre-lationships between water and energy usage in watershed scale and considering IWM.

2.2.3 Energy–Food

Considerations about the energy–food chapter in the food supply chain are principally related to the use of energy. Based on the degree of mechanization, agricultural products directly use energy as fuel for different tillage practices, crop and rangeland management, and transportation or electricity. It is also indirectly consumed through the use of energy inputs, such as fertilizers and pesticides or energy for the production of agricultural machinery [23]. By 2050, an increase of 80% and 60% is expected in the global need for energy and food, respectively. In order to reach an enhanced recognition about energy and food crisis resulted by developing countries, it is essential to concentrate on relationship between energy–food–water (EFW) and virtual water to connect the products trade, food, and water to each other [47]. According to Zhang and Vesselinov [95], energy is a critical element in food production and for mechanization, land preparation, fertilizer production and application, irrigation, packaging, food processing and storage. Another important dimension of energy–nexus is the influence of growing share of modern biology in the world's energy complex [23]. Portney et al. [65] investigated the food–energy–water (FEW) nexus awareness in the American public and assessed how that awareness was connected to policy support. According to their study, people awareness about energy–food nexus was at the least level and their awareness about water–energy nexus was at higher level. In addition, they argued that people with a higher level of awareness of the relationship between energy and food should be more supportive of energy efficiency, producing alternative energy to energy-intensive farming methods. In addition, the Energy–Food Awareness Index should be positively correlated with support for higher energy rates for high-volume users and higher licensing costs for wasteful restaurants, and perhaps to support more groundwater monitoring in hydraulic failure. From 2001 to 2016, Lu et al. [47] conducted a quantitative analysis in order to estimate the Central China region's energy–food production water footprint (WF) and virtual trade water flow. The outcomes illustrated the rise of the energy and food production WF and virtual trade water flow, which lead to water stress in local and export areas. In regard to this matter, recognition of different risks and unity of the energy–food nexus and also the effects on various stakeholders are predominant to follow energy and food security targets in tandem [23]. Risks and influences within the energy–food nexus have also been summarized in Table 4.

2.2.4 Water–Energy–Food

There is an inevitable connection among water, energy, climate and food security, and the associated natural resources. According to Machell et al. [50], the interdependency among water, energy, and food production is termed the water–energy–food nexus. Water, energy, and foods security (individually and across interconnections) illustrates that the WEF nexus required to be compiled and addressed in tandem to enhance our knowledge about the following matters [13]:

Table 4 Summary of risks and impacts within the energy–food nexus [23]

	Risks	Impacts
Energy-related risks to food security	Dependence on fossil fuels increases volatility of food prices and affects economic access to food	Fossil fuel dependency of upstream (e.g., production) and downstream (e.g., transport, storage, etc.) food supply chain Price volatility and supply shortages of energy inputs introducing economic and physical risks in the food supply chain Social, environmental and health impacts of traditional biomass cooking methods
	Potential trade-offs between bioenergy production and food crops	Allocation of agricultural products for bioenergy production with possible impacts on food prices
	Risks of energy production on food availability	Possible negative impacts of energy technologies (e.g., hydropower, thermal power generation) on river and marine life
Food-related risks to energy security	Overall increase in food production and changing diets raises energy demand along the food supply chain	Rising demand for energy needs for agriculture can strain the energy system, particularly in regions with a potential to expand irrigated agriculture with pumped water
	Quality and affordability of energy supply can depend on feedstock availability	In energy mixes with bioenergy, quality and affordability of foodcrop-based feedstock can affect energy supply

- The quality of the interconnection between the three parts,
- The effect of their changes and changes in other elements,
- The importance of policymaking and measures to address the three securities.

Water–energy–food nexus is an approach that coordinates management and controlling across parts and scales and follows enhanced recognition of the connection among water, energy, and food, both on resource utilization and also in international, legal and policy dimensions [37]. Treating the WEF nexus holistically in an ideal condition would resulted in more favorable resources allocation, enhanced economic adequacy, lower environmental and health effects and ultimately better economic development conditions [9]. The WEF chapters are so closely intertwined that each aspect is interdependent, ensuring access to services on issues of environmental, social, and ethical impact, as well as economic relations [64]. Various elements of the WEF nexus have been schematically shown in Fig. 5.

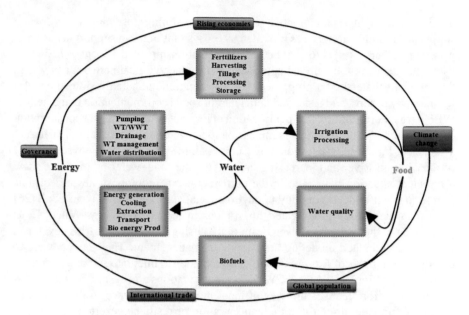

Fig. 5 Schematic illustration of various elements of the water–energy–food nexus (after [58])

Many studies have reported various attitudes of application on the WEF nexus, but few researches have been reported about the application of the WEF nexus at watershed scale. In this regard, a water–energy–nexus was application by Mayor et al. [53] for Duero River Basin, Spain. According to their investigations, some of the most critical subjects are limitations caused due to rise in energy prices for irrigated agriculture resulted from modernization, water treatment, and the other potential water demands for energy. A numbers of challenges exist within the WEF nexus at the local and global scales in agricultural land use in the watershed for food and biofuel productions. The first challenge is that high global energy prices cause to increase demand for bioenergy crops. It therefore leads to the reinforcement of completion for arable land and water with food crops. Another challenge is that the distinction of gender functions in trade and the synergy is often ignored [60]. In this respect, modeling the agricultural water–energy–food nexus in the Indus River Basin, Pakistan, has been studied by Yang et al. [91]. The result of their study indicated a growing trend in using water and energy under the government of hotter climate. While more flexible water allocation policies can reduce the detrimental effects of climate change on agricultural water and energy use, leading to more crops and hydropower generation. Such policies may vary year to year due to increase the use of resources. Karabulut et al. [36] compared the benefits of supporting the –food–energy relationship by considering the environmental flow requirements for river ecosystems using the hydrological model of Soil and Water Assessment Tool (SWAT) in the Danube River Basin. They showed that modeling is crucial for conducting an integrated analysis of the ecosystem–water–food–energy nexus. It was further showed that spatial mapping is an effective tool for demonstrating the availability of water supply and monitoring services provided by ecosystems and can support nexus analysis. Spiegelberg et al. [76] studied the interrelationship of the local WEF nexus using the sustainable livelihood approach via a socioecological network analysis in the Dampalit Watershed, Philippines. The investigation indicated various life styles for the study groups, and it showed deficiency of social communication in the field of WEF nexus. Furthermore, the identification of indirect connections via consumption of the groups of food products is impossible. The water–food nexus in the Citarum Watershed, Indonesia, was studied by Lubis et al. [48]. Assessing the water quality of the Citarum River and increasing demand for the food–water relationship show a turbulent situation that, even in the current context, requires strategic decisions to reform water allocation policy to environmental sustainability. Givens et al. [26] in their in their study investigated Innovations in the Food–Energy–Water Nexus (INFEWS) in the Columbia River Basin (CRB) using different scenarios developed based on coordinated modeling and simulated management. Their targets were to determine a number of less commonly used methods to integrate social perceptions with the Food–Energy–Water (FEW) nexus investigations and to seek how this interdisciplinary effort changes the way of innovations and resilience in FEW systems are evaluated. They found that the concentration of resilience exerted on multiple relationships could unwittingly accentuate the condition imposed by the governing sector. The integration of social perspectives, which lighten issues of inequity, power, and social justice, can incorporate these deficits and notify future originations, and more care needs to be paid to social stimuli and outcomes. Yang et al. [90] examined the effect of climate change and human made alteration on the water, energy, food, and ecosystem divisions in the Niger River Basin, Nigeria. The results of their research recommended that dam construction can reduce the detrimental effects resulted from climate change on hydropower generation as well as on ecosystem wellbeing to some limit. Villamor et al. [81] used a framework of social metabolic process and energy flow investigation to assess the agro-ecosystem and land use change in the food and water–energy system by using for the Upper Snake River Basin (USBR), Idaho, USA. Despite the increase in arable land between 2002 and 2012 in the basin, a decrease in energy production for crop production was observed. On the other hand, the growth in industrial energy input for dairy industry showed that the watershed is a good example for bio-industrial system. The WEF nexus in the Upper Yellow River Basin was reported by Si et al. [74] using multi-objective optimization for reservoir system. The results showed that the recognition of the WEF profits is defined by proper operation of the Longyangxia Reservoir, and Longyangxia Reservoir should sustain a high-water level to guarantee the overall profits in the long term. Temporal investigation of indicators for water–energy nexus for hydropower plants and water pumping in the Lower Blue Nile Basin was reported by Basheer and Ahmed Elagib [8]. They found that increasing the height of the Reserve Dam in 2013 in the study area led to a significant change in annual energy production, days in power generation, daily energy production, water–energy efficiency, and annual energy for pumping. In addition, turbines with higher capacity could be installed in the Roseires Dam to make to use the most hydropower potential. A review of the effect of the governance structure of the Lower Colorado River Basin (LCRB) on the application of the FEW nexus concept was presented by Huckleberry and Potts [35]. Their results showed that changes in food production are controlled by the governance structure rather than the availability of water. In addition, in proportion to water, more energy was used for food and more energy for transporting water to cities was more than water for agricultural sector. Recently, Shi et al. [73] used the water–energy–energy–ecology nexus in a Bayesian network to analyze and manage water resources in the Syr Darya River Basin. Complete scenario analysis showed that, increasing the proportion of food products, upgrading the competence of water use in salt washing and irrigation and preventing damage in the short term is beneficial. In the long run, based on the increasing use of advanced drip irrigation technology from 50 to 80%, the annual inflow to the Aral Sea will increase significantly, reaching 6.4 km^3 and 9.6 km^3, respectively.

2.2.5 Initiative Soil–Water–Energy–Food Nexus

The well-knit soil-peace connection is managed by the demand for delimited, but necessary, elaborately lined to ecosystems services and performances. Availability of sufficient and alimentary food is necessary to the man welfare, peace and tranquility. Soil has influences on universal peace via its effects on the inquiry for food, which improves the relevancy of the Peak Soil concept. A greener revolution can be started through the rational governance of the soil and the environment [41]. Soil is a peace and security issue that is outlined in Fig. 6.

Soil is progressively considered as a core commodity in reinforcing the natural endowment and production based at external level. The soil health is crucial not only for food security but also for preparing many other ecosystem services. The conservation of soil and water at watershed scale interferences highlighting the demand for sustainable land management practices [69]. We need to maintain a specific health index for the soil so that it can be supposed to do what it can in the face of climate change, urbanization, aggravated agriculture, and disposal. We should change our soil knowledge in order to maintain its intended role [57]. According to Rasul and Sharma [67], sustainable agricultural practices created for preventing land degradation, save water and energy by increasing green and blue waters by reducing the use of high-consumption fertilizers. Significant successes have been achieved with the use of the new water management system [92] states that suitable treated wastewater, which can now be used for irrigation, can also be more efficient to the farmers without degrading soil properties. The supporting role of the soil system in water and food security is shown in Fig. 7.

On the other hand, both soil and water are considered to be among the most climate-vulnerable divisions between environmental sources [32]. Much cooperation exists among the various functions and processes, and that specific processes may contribute to different functions. For instance, the water cycling process contributes to the ecosystem function biogeochemical cycling, but when water cycling physically shapes the environment, it can also contribute to the habitat provision function.

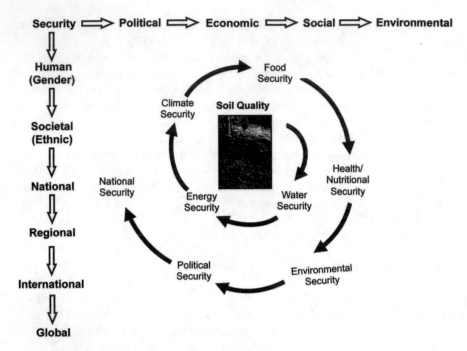

Fig. 6 Securitization of food and the environment through soil sustainability [41]

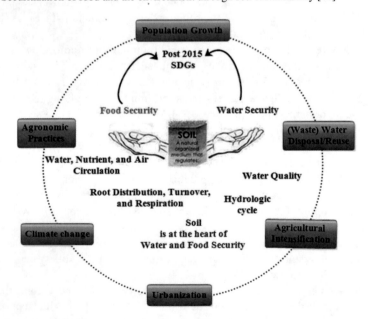

Fig. 7 Soil system at the heart of water and food security (after [57])

Functions such as soil fertility and primary production are highly dependent on the amount of nutrients in the soil/water, meaning they are closely connected to biogeochemical cycles [80]. In addition, reduction of land degradation can play a critical role in saving water and energy for human use through increasing soil water storage and groundwater recharge, and also decreasing the utilization of energy-intensive fertilizer [70]. The foundation of any food–water–energy nexus system is formed by the unique natural organization of soil. It also moderates the soil–water–plant–energy nexus with the replenishment of green-water supply (from precipitation) for plants and soil biota, which in turn enables the production of biomass as a resource of food, feed, fiber, and biofuel feedstock. Indeed, the soil is a very large reservoir for water and carbon with strong influences on local, regional, and global climate [43]. As Rasul [66] mentioned, the soil formation needs synergy among water, air, the parent materials of soil, and plants, whereas net ecosystem productivity needs soil, water, energy inputs, and living organisms. Besides, a strong nexus exists among population, food, climate, soil, water, and waste. The reduction in availability of natural resources (e.g., soil, water, and energy) and climate change lead to increase of the importance of such a nexus. With understanding this matter that larger challenges lie ahead, the crucial role of the nexus cannot be overemphasized [42]. The importance of soil as a nexus tool in addressing global issues is shown in Fig. 8.

There is a potent soil–water–waste nexus. Green water is the soil moisture from the blue water collected as precipitation and move on or through supplemental irrigation, impresses the biomass or net primary productivity (NPP). The surface runoff produced from soil provides to the blue water. Soil control causes crop and animal wastes that can be composted and utilized to improve/restore soil conditions. Likewise, gray/black water can be used for additional irrigation and as a resource of plant

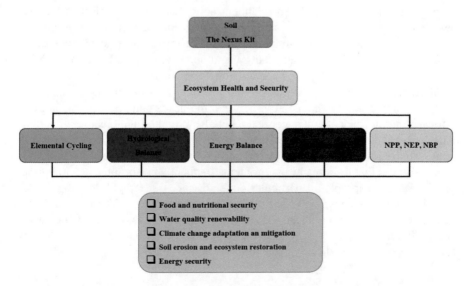

Fig. 8 Importance of soil as a nexus tool in addressing global issues (after [43])

Fig. 9 Conceptual diagram
of the role of soil in the food,
energy, and water nexus
(after [29])

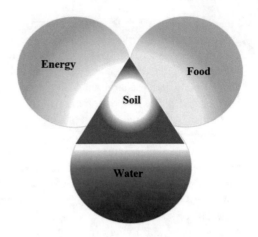

nutrients in agro-ecosystems [42]. Soil resources present basic ecosystem services, so are at the centers of all SDGs requirements. Nevertheless, soil sources are usually confidentially owned and are viewed as an issue of the national government and are difficult to contain in global sustainability programs [85]. Soil is the most over-looked part of the nexus between food, energy and water and accounts for 99% of the world's food. Nevertheless, there is oversight the nexus and it includes many of the environmental services that humans need. Water, energy, and food security describes the present and future problems of human sustainability while preserving the environment [29]. The role of soil in the food, energy, and water nexus is shown in a conceptual diagram in Fig. 9.

In this respect, Subramanian and Manjunatha [77] presented the concept of nexus among Energy–Water–Soil–Food in Indian agriculture. They confirmed misman-agement of these resources, in the long term, could build major warnings for future periods. Besides that, Mukuve and Fenner [59] used a nexus approach to study the impact of water, land, energy, and soil nutrient on the food system in Uganda. The benefits of this report are on food security strategy and resource connectivity for managing the present and future of Uganda's food scheme. Their findings pointed that the connection between the nexus components was not only intricate, interdependent and synergistic, but also posed limitations to productivity. Kirschke et al. [39] provided a detailed conceptual analysis of resource nexus prob-lems, using scrutinizing water–soil relationship on the Loess Plateau in China. The results showed that resource nexus problems frequently perform the general princi-ples of dangerous problems, particularly the presence of deeply different purposes, full system complexness, and high information uncertainness. The resource nexus strategy can finally discuss these gaps by knowing how the four systems (i.e., water, energy, food, and soil) combine to further equity and decision-making, also to unite sections of society and different scientific methods.

3 IWM and SWEF Nexus

The use of integrated watershed management (IWM) is now gaining more attention with the development of watershed management. IWM also provides basic watershed management ideas to combine different social, technical, and institutional dimensions, conservation, social, and economic goals [25, 82]. The IWM strategy illustrates the significance of viewing at various uses of watershed resources. It protects ecosystem services and biodiversity and tries to balance human and environmental requirements [7]. Wang et al. [82] described IWM as an adaptive, multicultural, and multidisciplinary systems framework control which proposes to store efficiency and ecosystem sincerity about the water, soil, plants, and animals across a watershed, through preserving and recovering ecosystem services for environmental, social, and economic benefit. Plans for IWM can have purely positive social effects on all characters of the water, energy, and food nexus [30]. Water, energy, ecosystems, and agriculture have always been recognized as the key sectors and related to an IWM approach. Nowadays, it is essential for all scientists and managers to discuss the changes occurred in the world and the regions, and consider water, energy, and food in an integrated manner [72]. In contrast, soil connects climate change, the deterioration in biodiversity, water, energy, and food security, and is a basic element of environmental sustainability problems [29]. To incorporate soil into the food, energy, water nexus, the concept of soil function should be developed to connect with ecosystem functions [2]. In this regard, Gebremeskel et al. [24] studied the impact of IWM on soil health in northern Ethiopia. They confirmed that a presentation of the IWM applications has positively modified soil health. Furthermore, IWM applications showed more effectiveness on soil health in comparison with other management practices performed. Teka et al. [79] also reported assessing the influence of IWM on decreasing soil erosion and shifts in the livelihoods of rural households in Ethiopia so that the rate of soil erosion was halved. In addition, increasing rates of crop productivity, availability of water for irrigation and domestic uses, availability of fodder and household income were 22, 33, 10, and 56%, respectively. IWM can be utilized as an approach point activity for developing livelihood for rural societies, an improvement in water availability, rehabilitation of ecological balance in the depraved rainfed ecosystems by greening these regions, and diversification of cropping farming systems [12]. Hence, the soil–water–energy–food (SWEF) nexus can be introduced as a useful framework for higher sustainability and adaptive management at the watershed scale.

4 Water–Energy–Food Nexus: Examples from Iran

A review of the literature on the water–food–energy nexus pointed that most studies reporting the WEF nexus from developed countries and limited cases from developing countries such as Iran. Few studies have been reported on application of WEF nexus in different regions of Iran. Mirzaei et al. [55] presented the study of the energy–food groundwater (GEF) nexus for Iranian agricultural products based on provincial and national data sets and first-hand approximations of agricultural groundwater exploitation. They reported that water consumption for crop production has significantly exceeded the country's renewable water supply capacity, posing a serious national security warning for water bankruptcy. In addition, the significant deterioration of groundwater levels in the country and the increase in energy consumption underscore the inefficient feedback links between agricultural water and energy prices and groundwater exploitation in an inefficient agricultural sector. In the same vein, a system dynamics model was developed by Bakhshianlamouki et al. [6] to quantify the impact of restoration measures on the water–energy–food nexus in the Urmia Lake Basin. They stated that a comprehensive restoration program could be effective in improving the lake level to the recommended ecological level by 2040. This study generally considers all parts of the nexus to assess the impact of all the measures introduced that appear to be positive on paper, but may have unintended consequences, such as an increase in the energy required by electric pumps. Then, dynamic modeling of the system was studied to assess the security of water, food, and energy resources (WFE) and nexus over a 10-year period in the Gavkhuni basin by Ravar et al. [68]. The results showed that the most effective methods to improve the status of the WFE system and meet the environmental demand of Gavkhuni Wetland were the related policies recommended by the agricultural and environmental sectors. Surface water security and groundwater security under the combined implementation of changing the product pattern and improving product efficiency and controlling groundwater exploitation programs, about 4 and 5%, respectively, but decreased water for food production and energy for water at respective tunes of 18% and 26%. Based on water–energy–food (WEF) Security Index reports, WEF Index for Iran has been found to be 0.68 [94]. A linear water–energy–food nexus optimization for planning 14 crops planted in orchard, irrigated farms, and rain-fed farms has been formulated for the Shazand Watershed, Markazi Province (Fig. 10), between 2006 and 2014 by Sadeghi et al. [71]. They followed five steps to perform the study as shown in Fig. 11.

In this regard, Sadeghi et al. [71] investigated water–energy–food (WEF) nexus framework as shown in Fig. 12. It combines three interlinked water–food components (i.e., water consumption, water mass productivity, and water economic productivity indicators), energy–food (i.e., energy consumption, energy mass productivity and energy economic productivity indicators), and energy–water (i.e., energy consumption indicator).

Then, Sadeghi et al. [71] obtained a spatial database of the Shazand watershed. Water–energy–food nexus index calculated for each crop. Finally, a linear optimization problem to maximize a water–energy–food nexus index (WEFNI) was introduced to get access to an optimal cropping pattern. The results pointed out that from 2009 to 2013, the maximum and minimum levels of water consumption were related to sugarcane and bean crops. Potato and onion had consumed the maximum energy in irrigated lands with respective amounts of 281,310 and 301,639 MJ ha^{-1} during three time spans of 2006 and 2010, and 2011 and 2014, respectively. Walnut, almond, and

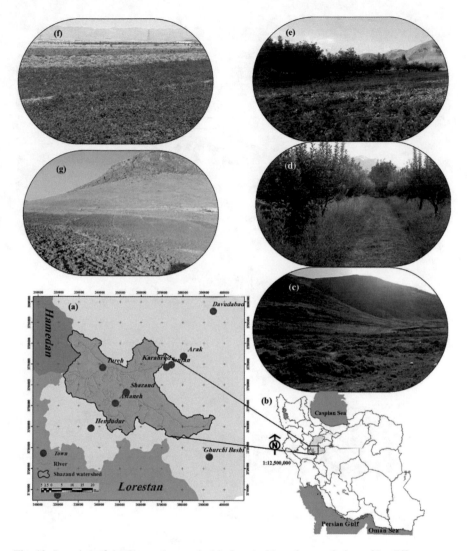

Fig. 10 Location of the Shazand watershed in Iran (**a–b**), and general view of its different parts (**c–g**)

wheat consumed the minimum energy in irrigated lands during the study periods of 2006 and 2009, 2007 and 2008, and 2010 and 2014, sequentially. Data analysis from 2006 to 2014 showed that the average direct and indirect energy consumption for irrigated and rain-fed crops in the Shazand Watershed was about 78,594 MJ ha^{-1}. The highest energy consumption and water productivity (cal. m^{-3}) were associated with Safflower and Almond. Walnut, watermelon, almond, barely also had higher energy productivity (cal. mj^{-1}) than other products in different years. The maximum water economic productivity (\$. m^{-3}) and energy economic productivity (\$. mj^{-1}) were

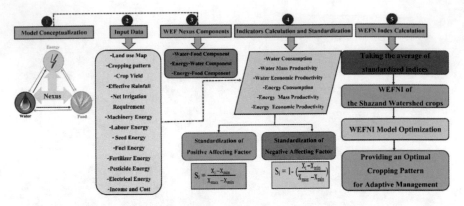

Fig. 11 A detailed flowchart for the water–energy–food (WEF) nexus index assessment

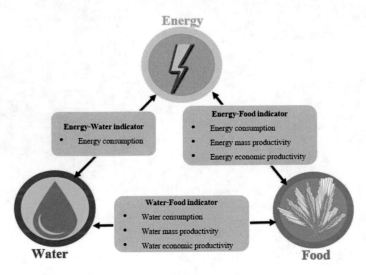

Fig. 12 General concept of the water–energy–food approach

consequently allotted to almond. Nevertheless, wheat had the lowest water economic productivity during the period between 2006 and 2008. Likewise, wheat and sugar cane had the lowest water economic productivity over the period of 2009–2014. While, wheat had the lowest energy economic productivity during 2006 and 2008. On the other hand, wheat and sugar cane had also the lowest energy economic productivity, over the study period. Ultimately, water–energy–food nexus index (WEFNI) was determined for the Shazand Watershed in Markazi Province, Iran, for eight representative irrigated farming crops, two representative orchard crops, and four representative rain-fed farming crops for the period of 2006–2014. WEFNI of the study crops has been shown in Table 5.

Table 5 Water–energy–food nexus index of the studied food crops in the Shazand watershed [71]

Crop	2006	2007	2008	2009	2010	2011	2012	2013	2014
Irrigated wheat	0.24	0.24	0.25	0.29	0.29	0.29	0.29	0.30	0.31
Bean	0.20	0.19	0.19	0.27	0.27	0.27	0.27	0.28	0.29
Sugar cane	0.18	0.17	0.17	0.11	0.11	0.11	0.11	0.10	0.16
Potatoes	0.05	0.05	0.05	0.06	0.09	0.10	0.11	0.11	0.14
Onion	0.19	0.18	0.18	0.27	0.35	0.31	0.31	0.29	0.30
Watermelon	0.33	0.32	0.34	0.39	0.30	0.30	0.30	0.30	0.31
Cucumber	0.25	0.25	0.25	0.19	0.19	0.20	0.22	0.23	0.13
Alfalfa	0.29	0.14	0.15	0.25	0.25	0.24	0.25	0.25	0.26
Almond	0.92	0.76	0.76	0.83	0.86	0.86	0.87	0.87	0.88
Walnut	0.59	0.42	0.41	0.50	0.52	0.52	0.52	0.52	0.53
Rain-fed wheat	0.37	0.48	0.35	0.36	0.34	0.35	0.35	0.37	0.38
Barley	0.39	0.39	0.41	0.42	0.37	0.39	0.41	0.45	0.47
Pea	0.36	0.35	0.36	0.37	0.35	0.35	0.35	0.36	0.36
Safflower	0.52	0.55	0.54	0.57	0.56	0.56	0.57	0.58	0.58

Finally, the results of optimization study verified that the present situation of land use allocation in the Shazand Watershed is undesirable in viewpoint of minimization of water and energy consumption and maximization of benefit. The results further pointed that the mean reduction benefit of water consumption with the help of nexus approach would be about 3 (or by 16%), 5 (25%), 6 (26%), 9 (44%), 0.004 (39%), 5 (26%), 6 (27%), 5 (28%), and 5 (28%) MCM, during 2006 and 2014. In the same vein, the mean reduction benefit of energy while using the nexus approach would be also about 13,046,608 (17%), 12,893,028 (13%), 12,660,667 (10%), 64,868,538 (53%), 14,351,215 (11%), 15,270,574 (11%), 15,316,001 (11%), 16,054,753 (11%), and 16,243,075 (11%) MJ during span periods between 2006 and 2014.

Accordingly, by virtue of changes considered in land use allocation through the nexus optimization approach, the total annual benefit of the study watershed would be ultimately about 62,109,086 (57%), 70,230,603 (54%), 60,526,585 (54%), 50,139,113 (47%), 133,833,833 (61%), 121,902,194 (55%), 86,939,365 (48%), 59,063,051 (42%), and 55,929,314 (40%) USD during the same period (i.e., 2006–2014). Given the access to fully subsidized water and energy in the agricultural sector in Iran and the inappropriate use of land in the Shazand Watershed, the thinking approach of the nexus can reduce poverty and create job opportunities among farmers and achieve sustainable development in the agricultural sector as the largest consumer of water through minimizing water and energy consumption and maximizing their efficiency.

5　Conclusion

This chapter took aim to stipulate a summary of the water, energy, and food nexus. Water, energy, and food are three highly connected systems. There is a need for a coherent knowledge base and database indicators and metrics that cover all relevant spatial and temporal scales and provisioning views. This nexus must be resolved in practice in a variety of physical and political environments, including in areas where little work has been done. Domestically, successful IWM experience, including the use of water–energy–food nexus approach to determine the optimal model of agricultural management at the watershed scale, should be studied for wider application. In fact, the problem-oriented nexus solutions can also be utilized locally. Such an improved linkage understanding can pave the way for new decisions and policies in a green economy. In this case, help and cooperation between the scientific community, stakeholders, and decision-makers is essential to address the complexity of resources management and the development of requests for WEF security. Besides, the concept of the water–energy–food nexus must be expanded, forming the soil–water–energy–food nexus. In fact, to address global issues (e.g., food insecurity, water scarcity, energy insecurity, and soil erosion), the response rests in the sustainable development of natural resources through the successful management of the soil, water, energy, and food nexus. Therefore, there is a need to create a framework that explicitly identifies connections between systems, reduces transactions, and creates synergies between sections. The WEF nexus should develop interdisciplinary, distinct, and obvious participation using clear and generally accepted approaches among all stakeholders. However, to reach more significant conclusions, more insight studies are needed in various aspects with the cooperation of different specialties in the diversity of spatial, temporal, and systemic scales. Besides, dynamic and comparative assessment of performance of adapting WEF as well as other derivate with considering more chapters is strongly advised to adaptively manage the watershed.

References

1. Abbott M, Bazilian M, Egel D, Willis HH (2017) Examining the food–energy–water and conflict nexus. Curr Opin Chem Eng 18:55–60
2. Adhikari K, Hartemink AE (2016) Linking soils to ecosystem services—a global review. Geoderma 262:101–111
3. Ahuja S (2015) Food, energy, and water. In: Food, energy, and water: the chemistry connection. Elsevier, 461 p
4. Al-Sumaiti AS, Banhidarah AK, Wescoat JL, Bamigbade AK, Nguyen HT (2020) Data collection surveys on the cornerstones of the water-energy nexus: a systematic overview. IEEE Access 8:93011–93027
5. Allan T, Keulertz M, Woertz E (2015) The water–food–energy nexus: an introduction to nexus concepts and some conceptual and operational problems. Int J Water Resour Dev 31:301–311
6. Bakhshianlamouki E, Masia S, Karimi P, van der Zaag P, Sušnik J (2019) A system dynamics model to quantify the impacts of restoration measures on the water-energy-food nexus in the Urmia Lake Basin, Iran. Sci Total Environ 134874

7. Bakker K (2012) Water security: research challenges and opportunities. Science 337:914–915
8. Basheer M, Ahmed Elagib N (2019) Temporal analysis of water-energy nexus indicators for hydropower generation and water pumping in the Lower Blue Nile Basin. J Hydrol 578:124085
9. Bazilian M, Rogner H, Howells M, Hermann S, Arent D, Gielen D, Steduto P, Mueller A, Komor P, Tol RSJ, Yumkella KK (2011) Considering the energy, water and food nexus: towards an integrated modelling approach. Energ Policy 39:7896–7906
10. Benson D, Gain AK, Rouillard JJ (2015) Water governance in a comparative perspective: from IWRM to a 'nexus' approach? Water Altern 81:756–773
11. Bhaduri A, Ringler C, Dombrowski I, Mohtar R, Scheumann W (2015) Sustainability in the water–energy–food nexus. Water Int 40:723–732
12. Bhan S (2013) Land degradation and integrated watershed management in India. Int Soil Water Conserv Res 1:49–57
13. Bizikova L, Roy D, Swanson D, Venema HD, McCandless M (2013) The water-energy-food security nexus: towards a practical planning and decision-support framework for land-scape investment and risk management. International Institute for Sustainable Development, Winnipeg, pp 16–20
14. Brears RC (2015) The circular economy and the water-food nexus. Future Food J Food Agric Soc 3:53–59
15. Cairns R, Krzywoszynska A (2016) Anatomy of a buzzword: the emergence of 'the water-energy-food nexus' in UK natural resource debates. Environ Sci Policy 64:164–170
16. Daher BT, Mohtar RH (2015) Water–energy–food (WEF) ncxus tool 2.0: guiding integrative resource planning and decision-making. Water Int 40:748–771
17. Dai J, Wu S, Han G, Weinberg J, Xie X, Wu X, Song X, Jia B, Xue W, Yang Q (2018) Water-energy nexus: a review of methods and tools for macro-assessment. Appl Energy 210:393–408
18. de Andrade Guerra JBSO, Berchin II, Garcia J, da Silva Neiva S, Jonck AV, Faraco RA, de Amorim WS, Ribeiro JMP (2020) A literature-based study on the water–energy–food nexus for sustainable development. Stoch Environ Res Risk Assess 6:1–22
19. Ding T, Liang L, Zhou K, Yang M, Wei Y (2020) Water-energy nexus: the origin, development and prospect. Ecol Model 419:108943
20. Enerdata (2018) World energy statistics Enerdata. Available https://yearbook.enerdata.net. Accessed 29 Dec 2019
21. Fan JL, Kong LS, Wang H, Zhang X (2019) A water-energy nexus review from the perspective of urban metabolism. Ecol Model 392:128–136
22. Fang D, Chen B (2017) Linkage analysis for the water–energy nexus of city. Appl Energy 189:770–779
23. Ferroukhi R, Nagpal D, Lopez-Peña A, Hodges T, Mohtar RH, Daher B, Mohtar S, Keulertz M (2015) Renewable energy in the water, energy and food nexus. IRENA, Abu Dhabi, pp 1–125
24. Gebremeskel K, Teka K, Birhane E, Negash E (2019) The role of integrated watershed management on soil-health in northern Ethiopia. Acta Agric Scand Sect B Soil Plant Sci 69:667–673
25. German L, Mansoor H, Alemu G, Mazengia W, Amede T, Stroud A (2007) Participatory integrated watershed management: evolution of concepts and methods in an ecoregional program of the eastern African highlands. Agric Syst 94:189–204
26. Givens JE, Padowski J, Guzman CD, Malek K, Witinok-Huber R, Cosens B, Briscoe M, Boll J, Adam J (2018) Incorporating social system dynamics in the Columbia River Basin: food-energy-water resilience and sustainability modeling in the Yakima River Basin. Front Environ Sci 6:1–19
27. Granit J, Jägerskog A, Lindström A, Björklund G, Bullock A, Löfgren R, de Gooijer G, Petti-grew S (2012) Regional options for addressing the water, energy and food nexus in Central Asia and the Aral Sea Basin. Int J Water Resour Dev 28:419–432
28. Hagemann N, Kirschke S (2017) Key issues of interdisciplinary NEXUS governance analyses: lessons learned from research on integrated water resources management. Resour 6:9. https://doi.org/10.3390/resources6010009

29. Hatfield JL, Sauer TJ, Cruse RM (2017) Soil: the forgotten piece of the water, food, energy nexus. In: Advances in agronomy, vol 143. Academic Press, pp 1–46

30. Haynes J (2012) Routledge handbook of democratization. Routledge, 470 p

31. Healy RW, Alley WM, Engle MA, McMahon PB, Bales JD (2015) The water-energy nexus: an earth science perspective. U.S. geological survey circular 1407, 107 p

32. Hettiarachchi H, Ardakanian R (eds) (2016) Environmental resource management and the nexus approach: managing water, soil, and waste in the context of global change. Springer, 209 p

33. Hoff H (2011) Understanding the nexus: background paper for the Bonn 2011 conference: the water, energy and food security nexus, 122 p

34. Howarth C, Monasterolo I (2016) Understanding barriers to decision making in the UK energy-food-water nexus: the added value of interdisciplinary approaches. Environ Sci Policy 61:53–60

35. Huckleberry JK, Potts MD (2019) Constraints to implementing the food-energy-water nexus concept: governance in the Lower Colorado River Basin. Environ Sci Policy 92:289–298

36. Karabulut A, Egoh BN, Lanzanova D, Grizzetti B, Bidoglio G, Pagliero L, Bouraoui F, Aloe A, Reynaud A, Maes J, Vandecasteele I, Mubareka S (2016) Mapping water provisioning services to support the ecosystem-water-food-energy nexus in the Danube River Basin. Ecosyst Serv 17:278–292

37. Kattelus M, Rahaman MM, Varis O (2014) Myanmar under reform: emerging pressures on water, energy and food security. Nat Resour Forum 38:85–98

38. Khan S, Hanjra MA (2009) Footprints of water and energy inputs in food production—global perspectives. Food Policy 34:130–140

39. Kirschke S, Zhang L, Meyer K (2018) Decoding the wickedness of resource nexus problems—examples from water-soil nexus problems in China. Resour 7:67. https://doi.org/10.3390/resources7040067

40. Kurian M (2017) The water-energy-food nexus: trade-offs, thresholds and transdisciplinary approaches to sustainable development. Environ Sci Policy 68:97–106

41. Lal R (2015) The soil–peace nexus: our common future. Soil Sci Plant Nutr 61:566–578

42. Lal R (2013) Climate-strategic agriculture and the water-soil-waste nexus. J Plant Nutr Soil Sci 176:479–493

43. Lal R, Mohtar RH, Assi AT, Ray R, Baybil H, Jahn M (2017) Soil as a basic nexus tool: soils at the center of the food–energy–water nexus. Curr Sustain Energy Rep 4:117–129

44. Liu D, Guo S, Liu P, Xiong L, Zou H, Tian J, Zeng Y, Shen Y, Zhang J (2019) Optimisation of water-energy nexus based on its diagram in cascade reservoir system. J Hydrol 569:347–358

45. Liu J, Mooney H, Hull V, Davis SJ, Gaskell J, Hertel T, Lubchenco J, Seto KC, Gleick P, Kremen C, Li S (2015) Systems integration for global sustainability. Science 347

46. Liu J, Yang H, Cudennec C, Gain AK, Hoff H, Lawford R, Qi J, de Strasser L, Yillia PT, Zheng C (2017) Challenges in operationalizing the water-energy-food nexus. Hydrol Sci J 62:1714–1720

47. Lu S, Zhang X, Peng H, Skitmore M, Bai X, Zheng Z (2021) The energy-food-water nexus: water footprint of Henan-Hubei-Hunan in China. Renew Sustain Energy Rev 135:110417

48. Lubis RF, Delinom R, Martosuparno S, Bakti H (2018) Water-food nexus in Citarum watershed, Indonesia. IOP Conf Ser Earth Environ Sci 118(1):012023

49. Mabhaudhi T, Mpandeli S, Madhlopa A, Modi AT, Backeberg G, Nhamo L (2016) Southern Africa's water-energy nexus: towards regional integration and development. Water 8:235

50. Machell J, Prior K, Allan R, Andresen JM (2015) The water energy food nexus-challenges and emerging solutions. Environ Sci Water Res Technol 1:15–16

51. Martinez-Hernandez E, Leach M, Yang A (2017) Understanding water-energy-food and ecosystem interactions using the nexus simulation tool NexSym. Appl Energy 206:1009–1021

52. Mayor B, López-Gunn E, Hernaez O, Zugasti I (2015) The water-energy-food nexus: foresight for research and innovation in the context of climate change. European Commission, pp 1–208

53. Mayor B, López-Gunn E, Villarroya FI, Montero E (2015) Application of a water–energy–food nexus framework for the Duero River Basin in Spain. Water Int 40:791–808

54. McCarl BA, Yang Y, Srinivasan R, Pistikopoulos EN, Mohtar RH (2017) Data for WEF nexus analysis: a review of issues. Curr Sustain Energy Rep 4:137–143
55. Mirzaei A, Saghafian B, Mirchi A, Madani K (2019) The groundwater-energy-food nexus in Iran's agricultural sector: implications for water security. Water 11:1835
56. Moazeni F, Khazaei J, Pera Mendes JP (2020) Maximizing energy efficiency of islanded micro water-energy nexus using co-optimization of water demand and energy consumption. Appl Energy 266:114863
57. Mohtar RH (2015) Ven Te Chow Memorial Lecture: localizing water and food security. Water Int 40:559–567
58. Mohtar RH, Daher B (2012) Water, energy, and food: the ultimate nexus. In: Encyclopedia of agricultural, food, and biological engineering. CRC Press, Taylor and Francis Group, 2120 p
59. Mukuve FM, Fenner RA (2015) The influence of water, land, energy and soil-nutrient resource interactions on the food system in Uganda. Food Policy 51:24–37
60. Mwale JT, Mirzabaev A (2015) Agriculture, biofuels and watersheds in the waterenergy-food nexus: governance challenges at local and global scales. Change Adapt Soc Ecol Syst 2:91–93
61. Pacetti T, Lombardi L, Federici G (2015) Water-energy nexus: a case of biogas production from energy crops evaluated by Water Footprint and Life Cycle Assessment (LCA) methods. J Clean Prod 101:1–14
62. Pan SY, Snyder SW, Packman AI, Lin YJ, Chiang P-C (2018) Cooling water use in thermo-electric power generation and its associated challenges for addressing water-energy nexus. Water-Energy Nexus 1:26–41
63. Pereira-Cardenal SJ, Mo B, Gjelsvik A, Riegels ND, Arnbjerg-Nielsen K, Bauer-Gottwein P (2016) Joint optimization of regional water-power systems. Adv Water Resour 92:200–207
64. Peri M, Vandone D, Baldi L (2017) Volatility spillover between water, energy and food. Sustain 9:1–16
65. Portney KE, Hannibal B, Goldsmith C, McGee P, Liu X, Vedlitz A (2018) Awareness of the food–energy–water nexus and public policy support in the United States: public attitudes among the American people. Environ Behav 50:375–400
66. Rasul G (2014) Food, water, and energy security in South Asia: a nexus perspective from the Hindu Kush Himalayan region. Environ Sci Policy 39:35–48
67. Rasul G, Sharma B (2015) The nexus approach to water–energy–food security: an option for adaptation to climate change. Clim Policy 16:682–702
68. Ravar Z, Zahraie B, Sharifinejad A, Gozini H, Jafari S (2020) System dynamics modeling for assessment of water–food–energy resources security and nexus in Gavkhuni Basin in Iran. Ecol Indic 108:105682
69. Reddy VR, Syme GJ, Tallapragada C (2019) Module I: key features of sustainable watersheds. In: Integrated approaches to sustainable watershed management in xeric environments. Elsevier, pp 7–18
70. Rockström J, Falkenmark M, Allan T, Folke C, Gordon L, Jägerskog A, Kummu M, Lannerstad M, Meybeck M, Molden D, Postel S, Savenije HHG, Svedin U, Turton A, Varis O (2014) The unfolding water drama in the Anthropocene: towards a resilience-based perspective on water for global sustainability. Ecohydrology 7:1249–1261
71. Sadeghi SHR, Sharifi Moghadam E, Delavar M, Zarghami M (2020) Application of water-energy-food nexus approach for designating optimal agricultural management pattern at a watershed scale. Agric Water Manag 233:106071
72. Sharifi Moghadam E, Sadeghi SHR, Zarghami M, Delavar M (2019) Water-energy-food nexus as a new approach for watershed resources management: a review. Environ Resour Res 7:129–136
73. Shi H, Luo G, Zheng H, Chen C, Bai J, Liu T, Ochege FU, De Maeyer P (2020) Coupling the water-energy-food-ecology nexus into a Bayesian network for water resources analysis and management in the Syr Darya River Basin. J Hydrol 581:124387
74. Si Y, Li X, Yin D, Li T, Cai X, Wei J, Wang G (2019) Revealing the water-energy-food nexus in the Upper Yellow River Basin through multi-objective optimization for reservoir system. Sci Total Environ 682:1–18

75. Smidt SJ, Haacker EMK, Kendall AD, Deines JM, Pei L, Cotterman KA, Li H, Liu X, Basso B, Hyndman DW (2016) Complex water management in modern agriculture: trends in the water-energy-food nexus over the High Plains Aquifer. Sci Total Environ 566:988–1001

76. Spiegelberg M, Baltazar DE, Sarigumba MPE, Orencio PM, Hoshino S, Hashimoto S, Taniguchi M, Endo A (2017) Unfolding livelihood aspects of the water–energy–food nexus in the Dampalit watershed, Philippines. J Hydrol Reg Stud 11:53–68

77. Subramanian S, Manjunatha AV (2014) Demystifying the energy-water-soil-food nexus in Indian agriculture. Ecol Environ Conserv 20:303–312

78. Tan C, Erfani T, Erfani R (2017) Water for energy and food: a system modelling approach for Blue Nile River Basin. Environments 4:15. https://doi.org/10.3390/environments4010015

79. Teka K, Haftu M, Ostwald M, Cederberg C (2020) Can integrated watershed management reduce soil erosion and improve livelihoods? A study from northern Ethiopia. Int Soil Water Conserv Res 8:266–276

80. van den Heuvel L, Blicharska M, Masia S, Sušnik J, Teutschbein C (2020) Ecosystem services in the Swedish water-energy-food-land-climate nexus: anthropogenic pressures and physical interactions. Ecosyst Serv 44:101141

81. Villamor GB, Kliskey AD, Griffith DL, de Haro-Marti ME, Martinez AM, Alfaro M, Alessa L (2019) Landscape social-metabolism in food-energy-water systems: agricultural transformation of the Upper Snake River Basin. Sci Total Environ 135817

82. Wang G, Mang S, Cai H, Liu S, Zhang Z, Wang L, Innes JL (2016) Integrated watershed management: evolution, development and emerging trends. J For Res 27:967–994

83. Wang L, El-Gohary NM (2019) Understanding the water-energy nexus in urban areas: a cluster analysis of urban water and energy consumption. In: Proceedings, annual conference—Canadian society for civil engineering, June 2019, pp 1–10

84. Waughray D (2011) Water security: the water food energy climate nexus. In: World Economic Forum water initiative. Island Press. Washington, 272 p

85. Weigelt J, Janetschek H, Müller A, Töpfer K (2015) Editorial overview: environmental change issues: soils in the nexus. Curr Opin Environ Sustain 15:v–viii

86. Wijayanti Y, Anda M, Safitri L, Tarmadja S, Juliastuti, Setyandito O (2020) Water-energy nexus development for sustainable water management in Indonesia. IOP Conf Ser Earth Environ Sci 426

87. World Economic Forum Water Initiative (2012) Water security: the water-food-energy-climate nexus. Island Press, 272 p

88. Xiang X, Jia S (2019) China's water-energy nexus: assessment of water-related energy use. Resour Conserv Recycl 144:32–38

89. Xu Z, Li Y, Herzberger A, Chen X, Gong M, Kapsar K, Hovis C, Whyte J, Tang Y, Li Y, Liu J (2019) Interactive national virtual water-energy nexus networks. Sci Total Environ 673:128–135

90. Yang J, Yang YCE, Khan HF, Xie H, Ringler C, Ogilvie A, Seidou O, Djibo AG, van Weert F, Tharme R (2018) Quantifying the sustainability of water availability for the water-food-energy-ecosystem nexus in the Niger River Basin. Earth's Future 6:1292–1310

91. Yang YCE, Ringler C, Brown C, Mondal MAH (2016) Modeling the agricultural water–energy–food nexus in the Indus River Basin, Pakistan. J Water Resour Plann Manag 142:04016062

92. Yi J, Guo J, Ou M, Pueppke SG, Ou W, Tao Y, Qi J (2020) Sustainability assessment of the water-energy-food nexus in Jiangsu Province, China. Habitat Int 95:102094

93. Yumkella KK, Yillia PT (2015) Framing the water-energy nexus for the post-2015 development agenda. Aquat Procedia 5:8–12

94. Zarei M (2020) The water-energy-food nexus: a holistic approach for resource security in Iran, Iraq, and Turkey. Water-Energy Nexus 3:81–94

95. Zhang X, Vesselinov VV (2017) Integrated modeling approach for optimal management of water, energy and food security nexus. Adv Water Resour 101:1–10

The Use of Biodigesters in the Treatment of Swine Manure in Southern Brazil: An Analysis of an R&D Project from the Perspective of the WEF Nexus

Ruy de Castro Sobrosa Neto, Alexandro Luiz da Silva, Janayna Sobrosa Maia, Nei Antonio Nunes, Jacir Leonir Casagrande, and José Baltazar Salgueirinho Osório de Andrade Guerra

Abstract In 2008, the largest federal state company in the southern region of Brazil, Eletrosul, carried out the Alto Uruguai Project. This entailed 10 Canadian-type biodigesters, which were installed for the treatment of swine manure in Itapiranga County (located in the western region of Santa Catarina). In 2012, the Brazilian Electricity Regulatory Agency (ANEEL), through the Law 9.991/2000 (which promotes research and development initiatives in Brazil), launched a strategic R&D call. This targeted technical and commercial studies aspiring for the insertion of electricity generation from biogas (which arises from waste and liquid effluents in the Brazilian Energy Matrix). Eletrosul participated with a project that aimed to interconnect all the properties of the Santa Fé Baixa Line. The present was benefited by the Alto Uruguai Project, through a pipeline network and fiber-optic communication to a 480 kVA-installed capacity miniature thermoelectric power plant (MCT). Other large biodigesters were also built with the following materials: wood, concrete, stainless steel and slate stone. In this way, different technologies of biodigesters, and in different sizes, were built. This article aims to analyze, from the perspective of the water, energy and food nexus (WEF nexus), how the dimensions of sustainability can be met, providing opportunities for social, economic and environmental development for the region where the project was developed.

R. de Castro Sobrosa Neto · J. S. Maia (✉) · J. B. S. O. de Andrade Guerra
Center for Sustainable Development, University of Southern Santa Catarina (UNISUL), Rua Adolfo Melo 34, Florianópolis, Santa Catarina, Brazil

J. B. S. O. de Andrade Guerra
e-mail: baltazar.guerra@unisul.br

R. de Castro Sobrosa Neto · A. L. da Silva
CGT Eletrosul, Rua Deputado Antônio Edu Vieira, Florianópolis, Santa Catarina 999, Brazil

N. A. Nunes · J. L. Casagrande
University of Southern Santa Catarina (UNISUL), Rua Adolfo Melo 34, Florianópolis, Santa Catarina, Brazil
e-mail: nei.nunes@unisul.br

J. L. Casagrande
e-mail: jacir.casagrande@unisul.br

Keywords WEF nexus · Biodigester · Manure · Swine · Social innovation · Environmental impact · Treatment strategies · Bioresource

1 Introduction

The water, energy and food nexus emerged at the Bonn Conference in 2011 [38], therefore after the Alto Uruguai Project, which started in 2008. However, Eletrosul has since been working on a project that aims, from the treatment of swine manure (food), to generate electricity. By using this waste, instead of simply releasing it *in natura* directly in the Uruguay River Basin, the environmental impact is minimized, demonstrating social responsibility and leadership in alignment with the best global sustainability practices. Al-Saidi and Elagib [4] in their work to understand the integrative approach of WEF nexus pointed out that it is currently quite popular in environmental management, and that the concept has found fertile ground in science and policymaking. However, the authors also mentioned that there is no consistent view regarding the meaning of integration in the nexus, and that, while it has been quite successful in changing policy debates, the nexus lacks prioritization of issues and appears to be left to specific case studies and policymakers' choices. Finally, they highlighted that governance is the missing link in the debate on the topic. In this sense, a federal state company in the energy sector signing a cooperation agreement with a pig farmers' association, therefore implementing a business model that can be identified with the recommendations of the nexus, as well as with the elaboration of a business model to producers, demonstrates that Eletrosul contemplated through its research the formulation of governance-oriented policies.

The state of Santa Catarina is the penultimate southernmost state in Brazil, and its western region stands out for the production of poultry and pigs. According to Santa Catarina State Government, agriculture accounted for 6.1% of the gross value added of the state in 2017 and has had a growth of 9.5% [33]. In the far west of Santa Catarina is located the Itapiranga County, on the banks of the Uruguay River, bordering Rio Grande do Sul (RS) and Argentina. In the Tupi–Guarani indigenous language, Itapiranga means "red stone," which is a consequence of the basaltic soil that prevails in this region, and since the native peoples of Itapiranga were mostly Tupi–Guarani and Caboclo Indians, having had little contact with the colonizers, which in this case are mostly German [57].

Itapiranga stands out economically for its pig farming, and as a result of this production it was found that animal waste was polluting the Uruguay River Basin [22, 23]. So, the breeders, together with county administrators and local leaders, requested Eletrosul's support. Thereby, through research and development (R&D) initiatives, a solution could be sought for this situation, and thus, in 2008, the Alto Uruguai Project was implemented. It entailed 10 Canadian-type biodigesters (covered lagoon), which were installed with the forethought of building a miniature thermoelectric power plant (MCT). This would be supplied by the production units in order to convert biogas into electricity, enabling then, other improvements in the business model currently installed on the site [26, 68].

The Alto Uruguai Project was not completed as initially planned, being limited to the construction of 10 (ten) biodigesters in Itapiranga, which allowed the treatment of waste with the resulting biogas flaring. This was, however, without the monetary gain that electricity generation from biogas enables. In such a manner that when analyzing the dimensions of sustainability, the environmental dimension was taken into account, once the waste stopped being released without treatment in the Uruguay River Basin; however, both the economic and social dimensions were not addressed. The most well-known definition of sustainable development is that one must seek to meet the needs of the present generation without compromising the ability of future generations to meet their own needs. In this sense, Herremans and Reid [35] believe that some people may not agree that the conflicts discussed are really conflicts, and that when debating in search of solutions, opinions often differ as to what actions must be taken in order to achieve sustainability goals.

In 2012, Eletrosul participated, through its R&D area, in a public call for ANEEL Strategic R&D (number 014/2012). It was denominated Strategic Project: "Technical and commercial arrangements for inserting electric energy generation from waste-and-liquid-effluent biogas in Brazilian energy matrix." This project was launched by the Brazilian Electricity Regulatory Agency (ANEEL), which is the regulatory agency for the Brazilian electricity sector [1]. Its source of resources was based on Law 9.991/2000, which provides for investments in research and development by concessionaires, licensees and authorized companies in the electric energy sector (Site 2020). The third section of this law states that: "The concessionaires of electricity-transmission public services are obliged to invest, annually, the amount of, at least, one percent of their net operating revenue in research and development of the electric sector." In addition to that, from this 1% of net operating revenue, 40% (forty percent) must be allocated to research and development projects, according to regulations established by ANEEL. To simplify, throughout this article the second Strategic Project will be called "Biogas Project," while the first project carried out in the region will be called "Alto Uruguay Project."

In this second stage of Eletrosul's R&D actions in Itapiranga, a gas pipeline and the construction of other biodigesters, but on an industrial scale, were planned. They were manufactured from different materials in order to evaluate possible cost-effective relationships in power generation from swine manure, as well as to assess the costs of manufacturing and maintaining biodigesters. So that, in this way, Eletrosul would signal actions to encourage biodigester construction, through its R&D area, associating research and specialized institutions in order to investigate the best constructive strategies. Similar initiatives have already been carried out in Cambodia, Zimbabwe, Kenya, Tanzania and Uganda [40, 42] (Clemens et al. 2018). However, even though Itapiranga has a lower number of biodigesters installed, different construction methods were prioritized, with the interconnection of all units through a gas pipeline and fiber-optic network to a miniature thermoelectric power plant (MCT).

Of the 10 (ten) biodigesters initially installed in the Alto Uruguai Project, 2 (two) were deactivated and that is why the pipeline did not contemplate the connection of these properties, once the producers gave up on the pig farming activity. Another action taken by Eletrosul in this second phase was to contract the geomembrane exchange for the 8 (eight) Canadian-type biodigesters, also known as covered lagoon. This is due to the fact that the initially thought business model should provide a revenue source for producers through electricity generation from the energy injection generated in the electricity grid of the local electricity distribution company, Centrais Elétricas de Santa Catarina S.A. (Celesc, [16]). However, as the Alto Uruguay Project did not install the generating unit (in this case a MCT), Eletrosul (in conjunction with the Biogas Project) made it possible to replace these biodigesters geomembranes, which were in use for more than 10 (ten) years, exceeding the manufacturers' warranty and being unable to continue operating [5, 55]. The Biogas Project also aims to provide the farmers with a guiding material of the business model which their association must adhere. The business model must foresee operating and maintenance costs, including the next geomembrane exchanges, as well as performing the operation and maintenance of MCT and other facilities. In this manner, farmers can proceed independently, with their own resources, achieving the growth and expansion of their businesses.

Eletrosul is a federal state-owned company in the Brazilian electrical sector, an Eletrobras System's subsidiary, which operates in Mato Grosso do Sul state, as well as in the three southern states: Santa Catarina, Rio Grande do Sul and Paraná. On January 2, 2020, Eletrosul was incorporated by the company CGTEE, located in Rio Grande do Sul state and also an Eletrobras System's subsidiary. From the unification of these two companies, the Electricity Generation and Transmission Company of Southern Brazil (CGT Eletrosul) was born. This information is crucial, given that from 2008 until January 2, 2020, the company which led the present R&D project was Eletrosul, and it continues under the management of the new company, CGT Eletrosul. Having made the clarification about this merging process, throughout this article no distinction will be made between the companies Eletrosul and CGT Eletrosul, using the nomenclature Eletrosul to refer to both [18]. This chapter aims to analyze, from the perspective of the water, energy and food nexus (WEF nexus), how the dimensions of sustainability can be met, providing opportunities for social, economic and environmental development for the region where the project was developed.

2   The Social Innovation of the Research

Historically, governments have had difficulties in satisfactorily meeting the presented social demands. This indicates that management models used in the last decades need to be reassessed. In this sense, several movements seek to minimize the negative effects of this reality. New solutions need to be devised, in order to overcome the postmodern society challenges with innovation, and their complex relationships and interactions [9] (Cloutier 2003; Mulgan et al. 2007; Brunstein et al. 2008; Gregoire

2017). In this sense, Feyerabend [29], little used to epistemological conservationism, believed that through iconoclastic and innovative practices, science could contribute to human development. Especially because any "rule" (such as knowledge or social experience), even if seemingly "fundamental," should not become a canon, limiting knowledge evolution and social transformation and improvement.

In this perspective, a methodological pluralism thesis for different researches defended by Feyerabend finds support in German physicist Albert Einstein, who defined insanity as: "to keep doing the same thing and to expect different results." Recalling that, for nineteenth-century positivism, scientific knowledge should be restricted to observable phenomena scrutiny, as well as to experience extracted facts, marking a conviction by Auguste Comte that science is not entirely objective [19, 72]. Far from the scientific model of the nineteenth century and its reflexes that reach the present day, social innovation stands out as knowledge and experience aimed at solving social matters at various levels and challenges imposed by the constantly changing market. Therefore, it can be seen that in the face of social and economic demands, social innovation is in constant construction and improvement (Rodrigues 2007; Varadarajan 2014; Manzini 2014; Rao-Nicholson et al. 2017; Van Wijk et al. 2019).

The attitude of proposing paradigmatic changes that make unprecedented world-views with an innovative impact on society is not new. Kuhn [45], it is worth mentioning, observed that science is divided into two conceptions, the formalist and the historicist. In his work "The Structure of Scientific Revolutions," he presents a revolution in the reflection on science. That is, the change in knowledge may be the result of an innovative paradigmatic change.

Specifically on innovation, with the participation of energy debate in the inter-disciplinary field, Hess and Sovacool [36] in their research (with an initial body of 262 journal articles and books with a stratified sample of 68 published from 2009 to mid-2019) noted that theoretical frameworks associated with Science and Technology Studies are becoming increasingly prominent in energy research within social sciences. As an interdisciplinary field, Science and Technology Studies have no easily demarcated boundaries. This is made clear by demonstrating that students find, in educational programs around the world, the terms "Science, Technology and Society" and "Science and Technology Studies" being oftenly used in an equivalent manner.

Despite the whole process of crystallized knowledge specialization in methodolo-gies and investigative spectra, since the beginning of modernity, there were dissonant voices that proposed an interface between knowledge and practices in an innovative manner. For Martins and Monteiro [52], throughout his life as a scientist, John Locke (1632–1704) worked with different branches of knowledge, such as: epistemology, education, politics and ethics. His work "An Essay Concerning Human Understand-ing" resulted from the difficulties that arose in solving an epistemological problem, in which he suggested a previous question about the extent and limit of human understanding. It is worth saying that the path of sensitive experience as an answer to his questions allowed the development of a scientific methodology based on an unprecedented "protopsychology."

In addition to Locke, for Zanella et al. [72], among modern theorists with an outstanding participation in the debate between rationalist and empiricist approaches, names like Francis Bacon, René Descartes and Immanuel Kant stand out. The empiricist Bacon considers that there are two ways of generating and spreading knowledge doctrines: the "Anticipation of the Mind" and the "Interpretation of Nature." The first is for the cultivation of sciences, and the second, his choice of preference, standing for scientific discovery. The rationalist, Descartes, does not ignore the importance of empiricism in the formation of knowledge, but believes in the sieves of abstract reasoning and the deductive method. Such is given, once they avoid the mistakes created by human senses, not giving up mathematization and mechanism, because for Descartes the access to truth (philosophical or scientific) presupposes the use of a rigorous method (anchored in reason). The philosopher Immanuel Kant analyzed the limits and possibilities of knowledge, in his work "Critique of Pure Reason," emphasizing that knowledge does not immediately offer a sensitive experience [65]. The transcendental idealism proposed by Kant will influence the most different currents of social sciences in contemporary times. Among them are the social theories that support research and actions on innovations with an impact on current societies.

Joseph Schumpeter structured his economic thinking in studies that combine entrepreneurship and innovation. With innovation at the center of economic evolution, which can be defined as "the creation of a new production function," present in different dimensions of the production process. This employs the famous term "new combinations," which is frequently used in other roles developed in relation to "creative destruction" [21]. In the Schumpeterian perspective, three important stages mark technological innovation: invention, innovation and diffusion, that is, from the emergence of new ideas, through the development of products, processes and/or new services arising from this idea, until its insertion in the market. For Skarzynski and Gibson [63], profound cultural change requires time, money and commitment, while Kneller [43] observes that science pursues the truth and that technology preaches efficiency, so that science seeks to formulate the laws to which nature obeys. Technology uses these formulations to create implements and devices that make nature obey man. In this sense, it is important to highlight Broerse and de Cock Buning [14], when pointing that public involvement contributes to a more democratic form of normative decision-making and to a more needs-oriented innovation process, thereby increasing public appreciation.

In the appropriation of concepts belonging to the main theories and approaches to innovations and technologies resulting from human development, as well as their impact on the quality of life of societies that consume them. This project sought to understand the main trends and approaches of innovations and technologies applicable to the demands of ANEEL R&D, as part of the National Innovation System. Doing so through Eletrosul's debates with representatives of the community, where installation was predicted. In this sense, the role of the social actors involved on the development of innovations and technologies was discussed. In a manner that the agreements, terms of cooperation and contracts could be signed, allowing analyzes and comparisons of public innovation policies in Brazil with other countries.

This research can be understood as being a project that has inclusive economic, social and environmental development as a backdrop, seeking the convergence between innovation and technology. It took into account the role of the Brazilian Federal Government in the development of science, technology and innovation policies and systems (through a Federal Law to promote research and development). Such was performed in junction to universities (and its foundations), as well as with organizations of the civil society and for-profit organizations as axes of the development tripod in order to build knowledge. As a civil society organization, Eletrosul signed a cooperation agreement with the Bioenergy Association. Such was created through the union of local creators, and as for-profit organizations, there is the company contracted by Eletrosul that carried out the work in Itapiranga County, as well as its subcontractors.

Looking at categories, society and social innovations, as well as the interfaces between society, management and other modes of innovation, this project presented itself as a great challenge to Eletrosul. This is due to the fact that cultural, technical, political and social barriers had to be overcome. This shows how the concepts analytics, addressing the views on social ideals and evoking essential foundations in the understanding of the different models of organization and management of modernity and contemporaneity, were aligned. In this sense, the following were problematized: technological and market experiences; the social role of companies and the state; the actions of civil society organizations; the notion of social capital; and the types of projects in social innovation.

The project's social innovation stands out for the possibility of farmers to obtain a revenue from the generation of electric energy from biogas, resulting from the fermentation process of biodigesters. For that, all properties will have gas flow meters that will allow the accounting of the biogas volume produced in each property. Based on this value, as well as the sum of all producers, and the fact that the gas pipeline leads the biogas to the mini-thermoelectric plant, the participation percentage of each producer is accounted. For that, the concepts of innovation and society were crucial, considering a project that dealt with power relations, technologies and subjective practices, as well as corporate and social autonomy. The social and economic impacts of the implemented innovation models fostered (through a project aligned with the WEF nexus) innovative management, with development models aimed at the best market practices, in addition to the diffusion of innovation (technological and industrial), science and technology. The role of this research is remarkable, as it fostered the development of a specific law for the production, management, use and conversion of biogas (its characteristics, peculiarities, difficulties and demands), to assist public managers in updating the Biogas Law. Additionally, it also assisted the energy sector regulation in distributed generation models, which in turn can be an engine for social, economic and environmental developments, stimulating projects aligned with the water, energy and food nexus.

3   Applied Legislation

The legislation directly involved with this project is broad and diverse, involving and not being limited to: the promotion of R&D + I; actions related to distributed micro- and mini-generation; biogas usage and transportation; and hiring processes and acquisition of materials, equipments and services. In this sense, the characteristics of the Brazilian Electricity Sector R&D + I Law, the legislation related to distributed micro- and mini-generation, as well as the legislation related to biogas are presented below.

4   Characteristics of Both R&D Law and Innovation in the Brazilian Electricity Sector

On July 24, 2000, Federal Law No. 9,991 was enacted in Brazil. It provides for investments in research and development, as well as in energy efficiency, by concessionaires, permissionaires and electricity sectors authorized. This law determines that companies in the Brazilian electricity sector must invest 1% of their net operating revenue in R&D (from which, 20% of the amount is allocated to the maintenance of the Energy Research Company [EPE] of Brazil; 40% to the National Development Fund for Energy, Science and Technology [FNDCT] in Brazil; and another 40% in developing research and development projects aimed at improving the Brazilian electricity sector, with the purpose of giving back to Brazilian society through reasonable tariffs). In turn, Federal Law No. 13,203, of December 8, 2015, brought a change in Law 9,991/2000. It authorized companies linked to the Brazilian Mines and Energy Ministry (MME), as well as associates of the Electric Energy Research Center (Cepel) to apply, alternatively, investments destined to R&D projects in compliance with statutory obligations to provide institutional contributions to support and develop Cepel. This allocated resources that previously reached several educational, research and development institutions in national territory.

Cepel is considered a center of excellence in research aimed at improving the Electricity Power System (SEP), having a relevant role in the continuous improvement of the Brazilian National Interconnected System (SIN) throughout its more than 45 (forty-five) years of existence. However, R&D projects, according to the original intention of Law No. 9,991/2000, once destined to the modernization of the electrical system (as a way of meeting new demands and new realities that are presented with the technological development in such a close manner and made available society) end up meeting both price decrease and increase in energy availability. The Biogas Project is an unequivocal example of this condition, once when developed on an adequate scale, it is able to provide an electricity source at the consumption site itself. This reduces the demand for large transmission and generation structures, while minimizing installation's environmental impacts (with the interruption of improper wastes' release into the environment).

In addition to the mentioned laws, the Oslo Manual (which is a proposal for guidelines in data collection and interpretation on technological innovation, as well as being the printed and translated version in Brazil, was made by Studies and Projects's Financier [FINEP]) helps Brazil to guide and rank its research. This Manual distinguishes four types of innovation: product, process, marketing and organizational. It is based on: the demand for innovation indicators' consensus; the fundamental needs of policies and economic theory; the definitions and scope of innovation; and the lessons to be learned from other researches. Conducted jointly by OECD and Eurostat, the Oslo Manual was written for and by experts from around 30 countries that collect and analyze data on innovation. However, despite this manual observing that innovation occurs in different sectors of the economy, its approach prioritizes commercial relations. In order to observe social innovation, sources aimed at non-market-oriented processes should be sought.

Also in the spirit of encouraging research, Eletrobras Holding planned for the year 2020 the creation of the Innovation Award, aiming to reward its subsidiary companies and Cepel. It is a prize divided into three categories: (1) finalistic (generation and commercialization, transmission, services); (2) management and support; and (3) socio-environmental. Each of these categories is further subdivided into three modalities: (a) solutions from Eletrobras' R&D projects, as well as from generation and transmission companies; (b) solutions from initiatives developed by Eletrobras' employees, and generation and transmission companies; and, finally, (c) solutions from Cepel's R&D projects or initiatives developed by its employees. Due to the COVID-19 pandemic, the award implementation had to be postponed.

5 Characteristics of Brazilian Legislation Related to Distributed Micro- and Mini-generation

ANEEL Normative Resolution (REN) No. 482 of April 17, 2012, is the regulation which defines what is distributed micro- and mini-generation, as well as establishing the guidelines for network connection and the electric energy compensation system (considering that since its inception publication three other changes occurred, 2012, 2015 and 2017 [2]). In December 2019, a public hearing was held for its review, as established at the time of the change promoted in 2015. This REN initially presented a more generic configuration, where it established microgeneration as that of installed power up to 100 kW and mini-generation in the range between 100 kW and 1 MW (for sources based on hydraulic, solar, wind, biomass or qualified cogeneration). Also establishing the definition of electric energy compensation as a system in which the active energy generated by the consuming unit compensates the consumption of active electrical energy, according to the items I, II and III of article 2nd.

At first, REN was very limited as to the profile of the consumer who could claim connection to the system for compensation purposes. It established the deadlines for electricity concessionaires to adapt to the legislation and defined responsibilities for connection to the energy company and to the consumer. The first review, which took place in 2012, made an important change in the compensation definition for electricity. It stated that the electricity generated is provided free of charge through a loan to the local distributor and compensated in the future by the same consumer unit or by another with the same ownership of the previous one. It holds true whether it is an individual or a legal entity, regularly established and registered within the Federal Revenue Service.

The second review, in November 2015, brought more profound modifications. It changed the range of microgeneration to up to 75 kW, while creating two distinct categories of mini-generation of electric power: one ranging between 75 kW and 3 MW (for water sources), and another on for less than or equal to 5 MW (for qualified cogeneration or other renewable electricity sources). The 2015 review brought important news, which included: the project characterization with multiple consumer units, shared generation and remote self-consumption.

For REN No. 482/2012, enterprises with multiple consumer units are characterized by the use of electricity independently, in which each fraction constitutes a consumer unit. The service to areas of common use is separate consumer units, and all consumer units are in the same property or, at least, in contiguous properties, according to item VI of article 2nd. Shared generation, on the other hand, is characterized by the gathering of consumers, within the same concession or permission area of an electricity distributor (through a consortium or cooperative, composed of individuals or legal entities). They have a consumer unit with distributed micro- or mini-generation in a different place from the consumer units where the electricity will be compensated, according to item VII of article 2nd, while self-consumption is characterized by several consumer units with the same ownership as a legal entity, with headquarters and branch, as well as individuals who have a consumer unit in a different location from the distributed micro- or mini-generation unit (within the same concession area or permission where electricity is compensated).

As of the 2015 review, the term for offsetting credits generated by possible surpluses between energy production and consumption, which was 36 (thirty-six) months, has changed to 60 (sixty) months. Credits that have not been cleared within this period will be accounted for reversion in favor of the tariff moderation. The October 2017 review changed the definition of distributed mini-generation, removing water sources from those that can be characterized by concentrating generation on renewable energy sources. As established in article 15 of the 2015 review, until December 31, 2019, the resolution should undergo a new review. Although not yet published (until the writing of the article), the main point of the proposed revision is the application of charging for the costs of using the distribution system, hitherto exempt from compensation calculations. The justification presented by ANEEL is the correction of a growing imbalance in the remuneration of these costs to distribution companies that were being made only by consumers who did not have a distributed generation system framed in the forms of REN No. 482/2012.

6   Characteristics of Biogas-Related Legislation

Gas-related legislation appears in the Federal Constitution of Brazil of 1988, and its article 25, § 2nd states that it is up to the states to directly exploit or grant local piped gas services. In the State of Santa Catarina, the State Constitution of 1989 (in its article 8th, item VI) repeats the Federal Constitution of 1988 text, which determines the exploitation of the piped gas service. However, it was only in 1993 (as of the publication of State Law No. 8,999) that the creation of Santa Catarina Gas Company (SCGÁS) was established. It is responsible for meeting constitutional obligations with distribution exclusivity, according to article 2nd. In 1994, State Law 9,494 was published, defining the rules for the concession of public gas service channeled to SCGÁS. Then, on March 28, 1994, the Concession Contract was signed with exclusive gas distribution for the period of 50 (fifty) years, and in an exclusive manner to the concessionaire. However, there was no anticipation by the legislator for biogas, its production, transportation or use. Due to the lack of prediction, biogas is then treated as a gas (in general). This generic aspect brought an additional legal difficulty to the Biogas Project realization. Once, as stated in the aforementioned legislation, SCGÁS would be the only entity with autonomy to use public roads for gas transportation (with no possibility of delegation to third parties).

On July 4, 2018, after six months of negotiations (and with the consent of the Santa Catarina State Services Regulatory Agency [ARESC]), a Technical Cooperation Agreement was signed between Eletrosul and SCGÁS. It established the possibility of carrying out a R&D project in the State of Santa Catarina using public roads to install piped gas network. Soon after, on July 12, 2018, the State of Santa Catarina (through State Law No. 17,542) established the State Biogas Policy. It constitutes a broad set of definitions, principles, objectives, instruments, guidelines, regulatory aspects and from promotion to development of biogas (from the most diverse sources) production and use throughout Santa Catarina. The new law highlights: (i) its concern with balanced production within the scope of the water, energy and food nexus, since, as stated in its article 3rd, item I "the systemic vision of biomass and biodigestion management, which considers environmental, economic, cultural, social and technological variables" is the first of the legal principles that must guide all subsequent legislation, and (ii) clear goals aiming at the "protection of human and animal health and the environment," at the same time that allowing the reduction of stocks of "animal, urban and industrial wastes," in addition to "stimulating environmental labeling and consumption," according to items I, III and VII of its article 4th.

Within the scope of the research and development project, the new legislation, in addition to defining exactly what biogas is (thus differentiating it from gas in general), in its article 16 establishes that "the biogas transportation and distribution, through pipelines, do not equate to the distribution of piped natural gas." Thus, authorizing the use of public roads to drive the biogas from the production units to a gas recovery center, by converting it into electrical energy (as is the scope of the Biogas Project) or for heat generation or other type of usage (that producers organized in cooperatives or consortia can identify as advantageous). Consequently, in conjunction with ANEEL's REN No. 482/2012, biogas producers now have autonomy to achieve their goals. Once the institution of the State Biogas Policy, by modernizing legislation and providing new business opportunities to producers of biomass and biogas, offered sustainable development conditions based on food and energy production, respect for the environment and water preservation. All benefits are adequately accompanied by the responsibilities of those interested in adapting to the safety and quality constraints of the control bodies, as well as the resolutions issued under the protection of biogas and its usage.

7 Swine Production and Its Relationship with Water Footprint and WEF Nexus

Research and development projects that adopt the premises of WEF nexus tend to have a greater capacity to respond to society demands, such as: social, environmental and economic, as well as stimulating circular economy models. Lehmann [46] points out that a way to reduce energy consumption is to apply circular economy principles, and that this model seeks to base economic development on environmental and resource protection, so that the circular economy and the nexus are closely linked. The understandings of Chen et al. [17] that redesign, reduction, recovery, recycling and reuse (5R) practices for waste recovery, in addition to the implementation of WEF Nexus with green chemistry principles (GCP) can improve food security and resource sustainability. Freitas et al. [28] point out that biomass is a potential energy source for the diversification of Brazilian energy matrix. In this context, biogas produced from waste anaerobic digestion becomes a relevant renewable resource that may come to play an important role mitigating environmental problems and generating local electricity.

Animal protein production (as a food source) is the most complex form of food production. It is a rich nutrients' source, whose global demand will only grow in the coming decades, and from which the production complexity results in a high environmental impact. Livestock in Brazil is characterized by animals' division according to the stage of creation. In many times, breeders are cooperated by large companies that provide advice to its members, from animals' birth to their age of slaughter. These large companies search for animals on these properties, take them to their manufacturing units to slaughter, process animal protein and sell other associated products. In general, breeders specialize in stages such as creating, recreating or endings with little performance in slaughter, processing and commercialization (stages that add more value to the work done). In the case of the Biogas Project, breeders are divided into: (1) growth and termination unit; (2) initial growth unit; (3) weaner production unit; and (4) piglet production unit [50, 53, 62].

When thinking about water, energy and food, first there is water. In the current condition of the global population, estimated at more than 7 billion people by the UN [66], which points out that more than half a billion people do not have access to water (a number that is equivalent to the entire population of the Earth in the year of the discovery of Brazil by Portuguese colonists, 1500), as well as 2 billion people having access to water contaminated by fecal coliforms [34]. Marques et al. [51] question how to feed the world population without compromising present and future generations. Once the population and income changes of the coming decades tend to cause an increase in global demand for food, such as meat (its production being associated with environmental and public health impacts, as well as land and water depletion).

For Liu et al. [48], the water footprint measures the human appropriation of water resources for the consumptive use of surface and groundwater (blue) and soil water (green), and to assimilate polluted water (gray). In this sense, it is important to approach the amount of consumed water in livestock production and food in relation to the adopted processes, as well as the technologies used. According to this, Mekonnen and Hoekstra [54] present numbers that call the attention of those who do not know this reality. They present consumption studies of the three types of water: green, blue and gray; from the data presented in Table 1, we can analyze the water footprint in the production processes. The expression "water footprint" is a term that finds similarity and foundation in the previously created concept of ecological footprint [58]. It is a concept that allows to evaluate with more precision how countries establish their trade relations, thus bringing light to the analysis of trade flows from the perspective of such an important input as water.

Table 1 shows the water footprint resulting from protein production. Among the presented types (and observing the weighted average of water consumption per kilogram produced), it is found that pork production represents less than one-third of water consumption to be produced when compared to beef. In addition, it is also found that the quantities of water needed for pork production are equivalent to those of cheese production. For Xiong et al. [70], decoding the way in which final demand influences water consumption patterns is useful for understanding the discrepancy between water supply and demand. Wang et al. [69] highlight that agriculture is the basic unit that protects the world's food security and is also the sector that uses the largest amount of water in economic activities. Thus, combining the water, energy and food nexus with the water footprint concept allows the establishment of cost-benefit relations. So that, economic interests can be pursued, and natural resources can be used in favor of the local economy, in a sustainable manner. Consequently meeting the most well-known definition of sustainable development: seeking to meet the needs of the current generation without compromising the ability of future generations to meet their own needs [61]. Once, the water footprint allows estimating the water flows that are commercialized, whether in national or international trade [30].

The existing link between water, energy and food highlights the importance of sanitation. It also encourages the reuse of water in possible processes, helping to combat pollution of clean water and encouraging the use of polluted water as an energy source. Polluted waters are rich in carbon, an energy source for the application

Table 1 Green, blue and gray water footprint by types of animals and by-products for Brazil, China and India (liters) per kilogram produced

Animal products	Breeding system	Brazil			China			India		
		Green	Blue	Gray	Green	Blue	Gray	Green	Blue	Gray
Beef	Pasture	23,729	150	16	16,140	0	0	25,913	0	0
	Mixed	20,604	187	61	13,227	339	103	16,192	533	144
	Industrial	8421	147	244	10,922	933	1234	12,412	1471	866
	Weighted average	19,228	178	82	12,795	495	398	15,537	722	288
Pork	Pasture	5482	1689	318	11,134	205	738	3732	391	325
	Mixed	5109	828	316	5401	356	542	4068	893	390
	Industrial	8184	215	525	3477	538	925	9236	2014	1021
	Weighted average	6080	749	379	5050	405	648	5415	1191	554
Chicken	Pasture	6363	35	364	4695	448	1414	11,993	1536	1369
	Mixed	4073	32	233	3005	297	905	7676	995	876
	Industrial	3723	24	213	1940	195	584	3787	496	432
	Weighted average	4204	30	240	2836	281	854	6726	873	768
Eggs	Pasture	432	24	25	3952	375	1189	10,604	1360	1176
	Mixed	257	24	15	2351	230	708	6309	815	699
	Industrial	3625	28	213	2086	206	628	3611	472	400
	Weighted average	2737	27	161	2211	217	666	4888	635	542
Milk	Pasture	1046	22	7	1580	106	128	1185	105	34
	Mixed	1254	42	36	897	147	213	863	132	65
	Industrial	500	43	25	1488	76	56	444	61	100
	Weighted average	1149	33	22	927	145	210	885	130	63
Cheese	Pasture	5169	126	36	7812	540	633	5857	535	171
	Mixed	6197	225	178	4432	742	1055	4267	666	320
	Industrial	2471	227	124	7356	393	275	2196	315	493
	Weighted average	5681	178	107	4581	732	1036	4377	657	310

Source Adapted by the authors from [54] (2020)

of biodigesters. So, in the humanitarian context, the main links between energy and food security are pointed out as challenges in guaranteeing access to energy for approximately 3.1 billion people who live in low- and middle-income countries, especially in remote and off-grid areas (Caniato et al. 2017).

8 Biogas Generation Process from Swine Manure and the Energy Credit System

In order to provide a regular power generation capacity (as well as looking to benefit the largest possible number of rural producers and their families, thus covering a larger geographical area), the miniature thermoelectric power plant (MCT) idealized in the Alto Uruguai Project (and subsequently, designed and installed in the Biogas Project) is supplied by a total of 11 (eleven) biodigesters, from which 8 (eight) are remaining from the Alto Uruguai Project and 3 (three) are new ones built together with the MCT and interconnected through a piped gas network (gas pipeline). This is launched on rural roads with a total length of 11 (eleven) miles, and all the properties involved, in addition to the biogas pipeline, are connected through an optical fiber network launched in parallel to the gas pipeline. Such addition allows monitoring the operating conditions of each biodigester in real time.

The biogas generation process can be understood as having its initial stage based on factors associated with production animals, such as the diet choice, the breed of animals chosen for breeding and the age of the animals being raised (and from which the manure is being carried to the biodigester). These are factors that can result in more or less biogas production. The second stage is the washing of the floor where the animals are raised (and this waste is carried to the biodigesters and digestion ponds). In this stage, the biogas formation in the properties occurs, which is channeled and conducted through a gas pipeline to the MCT where the biogas is converted into electrical energy. Through the substation located next to the MCT, there is a connection to the electrical network of the region's energy distributor (Celesc). In this way, the energy injected into the electrical network becomes a credit for electricity to be consumed by the members of the cooperative (Fig. 1).

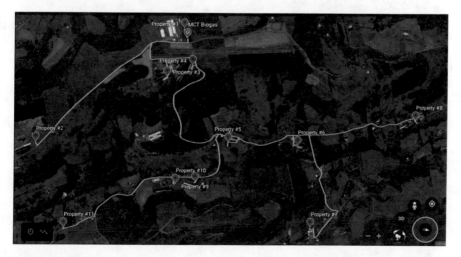

Fig. 1 Map of the gas pipeline layout. *Source* Elaborated by the authors, from Google Earth

The installed supervisory system is responsible for controlling the biogas production data of each property, analyzing, among other data, the quality and quantity of gas injected in the pipeline that reaches the MCT. Thus, from the data of all producers, it is possible to establish the proportion to which each producer is entitled with the generation of electric energy. For this calculation, the Biogas Project provided for a business model elaboration, which contains economic–financial support for producers to be able to manage the system after the completion of the R&D project conducted by Eletrosul. Owners will also need to forecast the costs of operating and maintaining the system, since it is a structure that combines: several biodigesters; gas pipeline; thermoelectric plant; substation and other components; and systems that require daily supervision and specialized labor. In addition to the Biogas Project facilities, some producers of the association already use complementary systems of solar collectors (on their own initiative) for washing breeding environments. Such is given since the city of Itapiranga has high thermal amplitude and water temperature is usually too low for direct contact with pigs.

9   Miniature Thermoelectric Power Plant (MCT)

MCT is responsible for converting biogas into electrical energy. In order to do this, the project has 4 (four) generator sets (totaling 480 kVA), sized for voltage levels of 380 and 220 V, and operating frequency of 60 Hz. The initially proposed generation offer suggested an operation of 8 (eight) hours a day, between 2 and 10 p.m., covering peak hours (the day interval between 5 and 10 p.m.). In this time interval, the Brazilian system meets the highest energy demand, for which consumers pay an additional value. In the same manner, energy production in the period is also remunerated in a different way (Fig. 2).

The choice of using 4 (four) generator groups was made by the team responsible for the generation research line, as well as to the supervisory system, which identified that only plant energy production systems relied on 1 (one) generator group. In addition to the quantitative, the issue of machine maintenance was considered. This allowed the construction of a preventive maintenance plan that can interrupt a machine, while

Fig. 2   MCT external and internal view. *Source* Pictures taken by the authors

the others remain operational. However, the equipment maintenance cost should be considered in the future evaluation (even though it is maintaining the production capacity of 360 kV). Once the maintenance costs are not restricted to the machine shutdown itself (with the loss of the remuneration resulting from it), but also include: displacement of authorized technician; replacement parts; tools; and materials and consumables. All of these are multiplied by 4 (four), even though the relationship is not direct according to the installed adaptation.

10 Substation

The substation is located next to the MCT, where (in addition to the transformer responsible for raising the voltage level) there is also the switching, protection and control equipments (such as disconnect switch, circuit breaker, and control and measurement panels). Its role is to receive the energy generated by the generator sets installed in the MCT and change its voltage level in the generation range (380/220 V at 60 Hz) to a value compatible with the Primary Distribution Network (RDP), which is provided by the electricity distributor that serves the region, Celesc. In the MCT, the RDP is 25 kV, at 60 Hz (Fig. 3).

The circuit breaker and the disconnect switch are protective equipments. They aim to isolate the transformer from the distribution network, once that for energy source isolation, only a circuit breaker is used. While the circuit breaker is capable of operating under the effect of electric voltage, as well as safely interrupting the current flow, the disconnect switch must be operated when the system is already interrupted (which prevents accidental system's restart). The control panels are the user interfaces for safely operating the equipment in the substation. Next to them is also installed the measurement system (for project-sized operations, and higher), which have memory resources capable of storing the quantitative of generation and consumption perceived in the plan. In such way that the concessionaire is able, with just a single measurement visit per month, to extract all necessary information necessary to verify values owed by, or owed to, the producer [20, 47].

Fig. 3 External and internal view of the electricity substation. *Source* Pictures taken by the authors

11 Gas Pipeline

In the Biogas Project, the set of materials and equipment (contained along its 11 km in length), from the biogas production unit to its storage at the MCT plan, was considered as a gas pipeline. It included: compressors, filters, gas dryers, polyvinyl chloride (PVC) tubing, maneuver and control valves, high-density polyethylene (HDPE) tubing, flare, etc. The compressors have the function of compressing the gas to be sent from the biodigester to the substation, overcoming differences in level and the distances that separate the production points and the point of conversion of the biogas into electrical energy, the MCT. Gas filters and dryers aim to remove impurities present in biogas, especially sulfur and water, which when combined accelerates the system corrosion as a whole. It is crucial to highlight that for biogas, the lower the presence of water and sulfur the better the performance and durability of the structure in its transport. Maneuver and control valves, in turn, are materials that allow directing the biogas flow. They are capable of isolating sections for safety and maintenance, directing the flow to prevent return to production plants, as well as isolating areas when necessary. The valves are installed in masonry boxes along the pipeline, more effectively at the connection points of the producers with the main network [8, 24, 32] (Fig. 4).

The basic structure for transporting biogas is made in HDPE piping, which must meet the Brazilian Regulatory Standard (NBR) No. 14.462. It needs to be orange in color and have the word "GAS" printed on its body. This is the only pipeline enabled for the transport of gas. The use of another pipeline type with another destination is liable to accelerated corrosion by biogas, representing a serious risk of accident when used. In addition to caring for safety in the system operation, it is also necessary to be careful with people who pass by. Even more so with any works that may be carried out close to the pipeline installation site. For this reason, signs were installed along the pipeline, with yellow plates and two different words: "CAUTION" and "BIOGAS NETWORK" [3, 7].

Fig. 4 Signpost and junction box with control registers. *Source* Pictures taken by the authors

Fig. 5 Example of a gasometer installed by Tecon Textile and the gasometer installation site of the Biogas Project in Itapiranga. *Source* Pictures taken from Tecon Textile's Web site and by the authors, respectively

12 Gasometer

The gasometer is a flexible structure made of HDPE, and in a double layer to meet safety issues, in order to provide greater mechanical resistance to the set. The volume of the Biogas Project's gasometer is 430 m^3. Its installation site is close to the MCT entrance, since it is stored in the biogas already purified to be used as fuel by the generator groups. It has a globe shape and is fixed to the base. It works as an energy storage system, operating as a control reserve that offers a continuous flow of biogas to generation. This minimizes the fluctuation effects in the production of biogas by biodigesters (Fig. 5).

Once installed, the gasometer must be cable-stayed (in addition to fixed to the base). This guarantees its safety and the stability of the system in cases of severe weather, as already verified at its installation site in Itapiranga.

13 Types of Biodigesters

In the Biogas Project, in addition to the installation of the MCT and the piped gas network, it also has the construction of 3 (three) biodigesters and 3 (three) wood digestate ponds. These are divided into: 1 (one) concrete digester; 1 (one) slate stone biodigester accompanied by 1 (one) wooden digestate pond; and 1 (one) stainless steel digester accompanied by 2 (two) wooden digestate ponds. Table 2 illustrates the types of biodigesters installed in the Biogas Project and their dimensions.

Table 3 shows the manufacturing price of the biodigesters and digestates installed in the Biogas Project, referenced on November 20, 2018. This was the date of the public call made by Eletrosul, and these were the values referring to the winning bid.

The slate stone biodigester, as well as the 3 (three) wooden digestate lagoons, has their outside structures reinforced with steel cable armor, in order to increase their mechanical resistance to the forces produced by the fluids inside. These steel cables

Table 2 Types of biodigesters installed in the Biogas Project of Itapiranga

Type of biodigester	Quantity	Dimensions in meters (diameter × height)	Capacity (l)
Slate stone	01	8.28 × 4.80	230,000
Stainless steel	01	11.49 × 4.80	680,000
Concrete	01	17.40 × 4.50	951,000
Wooden tank	01	8.56 × 4.50	230,000
Wooden tank	02	11.46 × 4.50	851,000

Source Elaborated by the authors

Table 3 Manufacturing price of biodigesters and digestates installed in the Biogas Project

Type of biodigester	Quantity	US dollar	Reais (Brazilian currency)
Slate stone	01	149,498.99	561,727.50
Stainless steel	01	565,363.55	2,124,297.00
Concrete	01	272,268.32	1,023,021.00
Wooden digestate	01	58,719.99	220,634.50
	02	298,598.50	1,121,954.00

Source Elaborated by the authors
Note The values of the dollar and real quotations were extracted from the Central Bank of Brazil Web site, www.bcb.gov.br/con versao

are from used elevator cables (which were taken out of use due to the end of their useful life for the transportation of people). However, by using a large number of cables around the tanks, they operate effectively for this function. Concrete and steel biodigesters will receive animal waste directly, while the slate stone digester and digestate ponds will have a vinyl wrapper inside (where the waste will be deposited). In this manner, they will behave as large containment structures for large vinyl bags, where the waste will be deposited, in addition to acting as a protection against ultraviolet rays, which cause the material to decompose.

13.1 Wooden Digestate Ponds

The wooden tanks are used as digestate ponds on the ground, as opposed to the traditional model of excavated ponds. The proposal of using tanks in this way offers the environmental advantage of perceiving the occurrence of leakage in the structure before infiltrating the soil (reduces harmful environmental impact). So, even though visually resembling biodigesters, they differ in terms of their function in waste treatment process. With regard to their construction method, they resemble large wooden barrels used for wine aging. The marked difference is that in the wooden barrels used

Fig. 6 View of the wooden digestates. *Source* Pictures taken by the authors

for the storage and aging of wine, the liquid is in contact with the wood. Such direct contact with the wood even has a chemical relationship in the wine aging (fermentation) process. In wood digestate, on the other hand, the liquid is not directly poured, as there is a large internal envelope where it is stored [71] (Fig. 6).

The wooden digestate ponds receive fluids from the digesters when the waste has already released its gaseous loads, remaining then liquid and solid components. In the digestate ponds, the material is deposited for decantation, so that the solid part is separated from the water. Subsequently, it goes through a press and drying process, in order to prepare for processing in a pyrolysis reactor, finally generating *biochar*, which is considered a high value-added fertilizer, free of pathogens and pollutants. The wooden digestate ponds were built on the properties where the slate stone and the stainless steel digesters are located. The digestate ponds were built with forestry wood, using autoclaved pine. The expectation is that their durability will be extended, once there is no direct contact with the waste (being subjected, then, mainly to mechanical stresses of the system and climatic conditions of the region). If the digestate ponds were built using native wood, the expectation is that their use would be greatly expanded and would minimize the chemical preservative application in wood treatment.

13.2 Slate Stone Biodigester

The slate stone digester was built from slate stone slabs extracted from a mine located 490 km from Itapiranga, in Trombudo Central County. Initially, it was believed that this raw material would be located closer to its application site, which did not occur in practice. In the slate stone slab production process, the initial prediction was for slabs to follow the dimensions: 10 cm × 50 cm × 480 cm (thickness × length × height). The ridges of the extraction site, however, pointed out that the stones extracted with these dimensions were too fragile for transportation (breaking by flexion when moved while still in the mine). Such event led to the slate stone blade thickness to be changed to 14 cm, increasing the weight of each slab by 300 kg (Fig. 7).

Fig. 7 View of the slate stone digester and broken slate stone slabs during transportation. *Source* Pictures taken by authors

For future projects, limitation for the slate stone slab height is recommended. This aspect influences its thickness and facilitates the cutting and transportations of the slabs, as well as the biodigester assembling.

13.3 Stainless Steel Biodigester

The stainless steel biodigester was built in slabs with dimensions of 300 cm × 100 cm, fixed with screws, washers and nuts. All of its structures were sealed with polyurethane, in order to waterproof the structure and prevent leakage from its interior during its retention time. One of the original proposals of the Public Call for Strategic R&D (from which the Biogas Project resulted) was the one of viable biodigester structures (considering the Brazilian reality) in order to enhance its insertion in the Brazilian energy matrix. The decision of choosing a steel biodigester (like the models disseminated on the European continent, such as Germany) was to study its viability in comparison with concrete structures, slate stone, covered pond and potential biodigesters to be built in reinforced wood steel, thus taking advantage of the digestate lagoon projects already developed in the project (Fig. 8).

With a 680 m³ capacity, it is able to receive and process the daily waste production from more than 2000 matrices, being the largest biodigester of the project. It is located in the same physical area as the MCT and the substation. Thus, it offers a guarantee

Fig. 8 View of the stainless steel biodigester. *Source* Pictures taken by the authors

Fig. 9 Frontal and superior view of the concrete biodigester. *Source* Pictures taken by the authors

of minimum generation operation even if the gas network becomes unavailable at some point.

13.4 Concrete Biodigester

The concrete biodigester was also designed as an alternative way of building biodigesters, adapted to the Brazilian rural context. However, it was verified, still in the construction phase, that this project needs to be reassessed in the future. Its assembly complexity is superior to the other biodigesters of the Biogas Project, especially with regard to the assembly of its box and hardware (steps prior to concreting). Its rounded shape and the necessary wall thickness made its assembly more difficult than planned, once the sliding shape technique could not be used (Fig. 9).

The difficulties identified in the assembly of the concrete biodigester demonstrate that its reproduction is not a trivial proposal. Especially if considering that the target audience are small and medium-sized producers, who are generally located in regions further from large centers and with limited resources. In such a manner that they would need to improve their production process in order to adapt to the increasingly complex and challenging Brazilian environmental legislation. As well as having to meet the growing consumers demands regarding environmental preservation. This is associated with what they willing to consume and pay a premium price when perceiving the producer alignment with their values [44].

13.5 Canadian-Type Biodigester or Covered Lagoon

The Canadian biodigester is a horizontal biodigester that has a rectangular base and a gasometer made of a flexible PVC blanket that surrounds the base. It has greater width than depth, resulting in a greater area of exposure to the sun, which contributes to the increase in biogas production (as the heat retention provides greater bacterial activity). This model is also known as covered lagoon, navy or balloon. It has low cost of implementation, easy maintenance and cleaning, with the disadvantage of having a short useful life, on average 5 years [5]. It was the model implemented by the Alto Uruguai Project, which proved to be very beneficial to producers due to the perception of improved working conditions and increased quality of life. The final report of the project, in June 2010, pointed to: "reduction of organic load and nutrients in effluents resulting from swine activity by up to 90% of the total emitted," in addition to "reduction in the number of dipterans in properties in Itapiranga, Riqueza, Guatambú, Águas de Chapecó, Palmitos and Pinheirinho do Vale" (which were the counties contemplated by the Alto Uruguai Project). At the time of the ANEEL Strategic R&D project, the installed biodigesters had already expired. Some of them still remained in operation, despite the weather action and ultraviolet solar radiation that accelerated the deterioration of the biodigesters still operational.

In order to guarantee the maximum production capacity for the Biogas Project, Eletrosul is providing the replacement of biodigesters depreciated by the action of time. The moment is propitious and meets a new reality verified in the Itapiranga region, which is the replacement of swine production for cattle in some properties. The biodigester for bovine production is basically the same for pig production, in its components and material composition. However, it keeps a difference in manure management, since the biodigester for application with cattle is (unlike the one designed for pigs) composed mainly of cellulose from the animal's feeding. While in swine production, all waste is washed and sent directly to the interior of the biodigester, in bovine production the solid material must be separated from the liquid portion. While the solid part is destined for composting, only the liquid is taken to the biodigester for gas production, resulting in a greater production of biogas.

14   Supervisory System

The supervisory system can be understood as an information traffic resource. It is increasingly widespread and associated with large energy transport systems that deal with safety-critical processes, such as electricity and gas. Its main function is to provide monitoring of complex and automated production processes, offering signaling and operation mechanisms, especially in cases requiring human intervention. For the Biogas Project, the supervisory system was designed with a tool to control the biogas production from each production unit, as well as its participation in the energy conversion process and its injection into the network, also allowing safety signals and guidance for monitored plant maintenance, in addition to forms of communication and operation, being then complemented by maintenance, operation, business model and communication. The supervisory system becomes, then, a fundamental element for the elaboration of the Biogas Project business model [32].

To carry out such monitoring, fiber-optic cables were also launched (together with the gas pipelines). They also connected the properties to the MCT, where a monitor offers a friendly interface for monitoring operational conditions. It is important to take into account the most important aspect for the constitution of a robust supervisory system: the safety in gas or biogas plans. In electrical systems (including large ones), even though their components are not mechanically covered by insulation elements, they are isolated by physical distance. A buried gas transport system deals with a highly combustible fluid chemical element. In case of leakage, it can travel great distances and accumulate in large areas before reaching the surface, which could ignite for any spark form, causing a large explosion. For this reason, supervisory systems must prioritize safety aspects, whether in control and operation of electrical, nuclear or thermal power plants, gas or biomass [25, 67]. It should be noted that the supervisory system of this project is not related to other traditional biogas production systems. It involves several participants and installations, distributed along 11 km of gas pipelines, in contrast to those in which production and conversion of biogas share the same physical and limited space to a producer (owner) only.

15   The COVID-19 Impact on the Final Stages of the Biogas Project

On February 6, 2020, Federal Law No. 13,979 was enacted in Brazil. It "provides for measures to deal with the public health emergency of international magnitude, resulted from the coronavirus outbreak in 2019." This law informed that the exercise and functioning of public services and essential activities should be safeguarded. On March 23, 2020, Santa Catarina State Government (where the Biogas Project work is situated) enacted Decree No. 525. It "provides for new measures to deal with the public health emergency of international magnitude, resulted from coronavirus, and establishes other measures." In this decree, the generation, transmission and distribution of electric energy and gas were considered public services and essential activities. However, the impact of COVID-19 on the final stages was due to the fact that some agencies adopted (totally or partially) the home–office model. In addition to that, Brazilian public universities interrupted their activities, and consequently, so did their foundations associated with research, development and innovation.

At the construction site, the greatest impact felt was on the requirement for social distance. In addition, other preventive measures, such as: mask wearing; washing hands frequently; and the already established regulations, regard health and safety of workers, as established by Brazilian regulatory standards. Hosseini [39] states that the COVID-19 pandemic has affected renewable energy manufacturing facilities, supply chains and companies, delaying the transition to the world of sustainable energy. The author further argues that several beneficial incentives should be offered by governments, in order to persuade private sectors, and society, to invest in renewable energy. Considering this, the Biogas Project is presented as a smart policy for converting animal waste (an environmental liability) into a sustainable solution, also being concluded in a moment of global pandemic, by a federal state-owned company in the Brazilian electric sector, and based on a Brazilian Federal Law to encourage research.

16 Final Considerations

When recapitulating important milestones to the Biogas Project, we are presented with: In 2000, the law to promote research in the Brazilian electric sector was created (Law 9,991/2000); in 2008, Eletrosul started the Alto Uruguai Project; in 2011, the Bonn Conference presented the concept from the WEF nexus. In 2012, Eletrosul participates, through its research and development area, in a public call for ANEEL Strategic R&D (number 014/2012). It was denominated "Strategic Project: Technical and commercial arrangements for inserting the generation of electricity from biogas from waste and liquid effluents in the Brazilian energy matrix," launched by ANEEL, the Biogas Project. Still in 2012, Normative Resolution (REN) No. 482 is elaborated. It defines what is distributed micro- and mini-generation, establishing guidelines for network connection and the electric energy compensation system. In this same year, its first review takes place. In 2015, the second revision of REN 482/2012 takes place. In 2017, the third revision of REN 482/2012 takes place. In 2018, after six months of debates and negotiations and with the consent of Aresc, a technical cooperation agreement is signed between Eletrosul and SC Gas. In January 2019, Eletrosul signs the contract that authorizes the beginning of the Biogas Project works in Itapiranga County, in response to the ANEEL R&D 014/2012. In February 2020, Federal Law No. 13,979 was enacted in Brazil, which "provides for measures to deal with the public health emergency of international magnitude, resulted from the coronavirus outbreak in 2019."

The existence of a law is to promote research and innovation in the Brazilian electric sector, from which its resource is under the management of companies that through their areas of research, development and innovation allow projects like this. It has social, environmental and economic reach, since it presents itself as a source of resources capable of fostering social innovation. This holds true not only in its place of implantation, but above all because it becomes a showcase of new concepts on both: sustainability; and water, energy and food nexus. This project presents itself as a national and international reference for presenting biodigesters of large and small sizes, and constructed from different techniques and materials, allowing to evaluate and identify the best cost-effective solutions to be applied in the treatment of swine manure, in addition to presenting a business model in which the local creators can carry out the system operation and maintenance, through the financial gains obtained with the generated energy, as well as other possible gains with biofertilizers, biochar and local tourism (fomented by visits to the site, in order to get to know the project).

Thus, the compliance with the objective of the electric sector R&D's law is evidenced. Once, this project demonstrated that the environmental, social and economic variables, of sustainability, are being met. Such is due to the fact that the waste treatment represented a great environmental gain, as well as a potential path for financial gains. In addition to these, the Biogas Project presented itself as a great showcase of concepts, enhancing social development beyond the producers' gains, by benefiting the entire community. The project's implementation (even though it started in 2008, about three years before the Bonn Conference) in its essence is structured in a relational logic with the principles of the water, energy and food nexus. The water is represented by the treatment of waste, which started to avoid pollution in the Uruguay River region. The energy is represented by the biodigesters, which operate from the swine waste. The food is represented by the pork meat, once animal protein is a food with greater production complexity (especially if observed from the perspective of water footprint).

Hence, the legislation's notion of encouraging research, development and innovation in the National Electric Sector (as in the case of Law 9,991/2000) is something to be celebrated. It imposes the challenge of change, by searching for efficiency and effectiveness, as well as overcoming old paradigms, with research, development and innovation as flagships, making a win-win system, society, the environment and the economic environment as a whole benefit from the achieved advances. Finally, it should be noted that this publication had the submission deadline of August 30, 2020. Therefore, the data contained in this publication (especially laws related to the described topic) must be analyzed until this date. Citing, as an example, that in August 2020, the Federal Chamber of Brazil's Deputies, through political party leaders, initiated actions to guide changes in the Gas Law (Law No. 11.909/2009). It introduces the concession regime. One of the major changes proposed in the New Gas Law project is the return of the authorization regime for gas pipeline transportation (coming closer to that practiced worldwide).

References

1. Agência Nacional de Energia Elétrica (ANEEL). Available in: https://www.aneel.gov.br. Accessed 12 Jul 2020
2. Agência Nacional de Energia Elétrica (ANEEL). Resolução Normativa n° 482, de 17 de abril de 2012. Available in: https://www2.aneel.gov.br/cedoc/ren2012482.pdf. Accessed 12 Jul 2020
3. Agência Nacional do Petróleo (ANP). Regulamento Técnico de Dutos Terrestres para Movimentação de Petróleo, Derivados e Gás Natural (RTDT). Available in: https://www.anp.gov. br/images/Legislacao/Resolucoes/2011/res_anp_6_2011_anexoI.pdf. Accessed 11 Aug 2020
4. Al-Saidi M, Elagib NA (2017) Towards understanding the integrative approach of the water, energy and food nexus. Sci Total Environ 574:1131–1139
5. Andrade MAN, Ranzi TJD, Muniz RN (2002) Biodigestores rurais no contexto da atual crise de energia elétrica brasileira e na perspectiva da sustentabilidade ambiental. In: Procedings of the 4th Encontro de Energia no Meio Rural
6. Arendt H (2013) The human condition. University of Chicago Press
7. Associação Brasileira de Normas Técnicas (ABNT). Available in: https://www.abnt.org.br/. Accessed 11 Aug 2020
8. Baredar P, Khare V, Nema S (2020) Design and optimization of biogas energy systems. In: Biogas digester plant, Chap 3. Academic Press, pp 79–155
9. Bauman Z (2000) Liquid modernity. Wiley Blackwell Publishing
10. Brasil. Lei 9.991, de 24 de julho de 2000. Dispõe sobre realização de investimentos em pesquisa e desenvolvimento e em eficiência energética... Available in: https://www.planalto.gov.br/cci vil_03/LEIS/L9991.htm. Accessed 14 Jun 2020

11. Brasil. Lei 13.203, de 8 de dezembro de 2015. Entre outras disposições dispõe sobre realização de investimentos em pesquisa e desenvolvimento... Available in: https://www.planalto.gov.br/ccivil_03/_Ato2015-2018/2015/Lei/L13203.htm. Accessed 12 Jul 2020

12. Brasil. Lei 13.979, de 6 de fevereiro de 2020. Dispõe sobre as medidas para enfrentamento da emergência de saúde pública de importância internacional decorrente do coronavírus... Available in: https://www.planalto.gov.br/ccivil_03/_ato2019-2022/2020/lei/l13979.htm. Accessed 25 Jul 2020

13. Brasil. Ministério de Minas e Energia (MME). Nova Lei do Gás. 2020. Available in: https://www.mme.gov.br/documents/36112/491926/Folder+Nova+Lei+do+Ga%CC%81s_julho_2020.pdf/5ee249d3-db91-b40d-aac1-a726a4a84997#:~:text=A%20Nova%20Lei%20do%20G%C3%A1s,modalidade%20de%20entradas%20e%20sa%C3%ADdas.&text=Mas%2C%20para%20que%20haja%20uma,de%20forma%20independente%20e%20coordenada. Accessed esso em: 12 Aug 2020

14. Broerse JE, de Cock Buning JT (2012) Public engagement in science and technology. In: Encyclopedia of applied ethics, 2nd edn, pp 674–684

15. Centrais Elétricas Brasileiras (Eletrobras). Available in: https://www.eletrobras.com. Accessed 12 Jul 2020

16. Centrais Elétricas de Santa Catarina S.A.—Celesc. Available in: https://www.celesc.com.br/. Accessed 12 Jul 2020

17. Chen TL, Kim H, Pan SY, Tseng PC, Lin YP, Chiang PC (2020) Implementation of green chemistry principles in circular economy system towards sustainable development goals: challenges and perspectives. Sci Total Environ 716:136998

18. Companhia de Geração e Transmissão de Energia Elétrica do Sul do Brasil (CGT Eletrosul). Available in: https://www.eletrosul.gov.br/. Accessed 12 Jul 2020

19. Comte A (1858) The positive philosophy of Auguste Comte. Blanchard

20. Costa JGDC, da Silva AML, Hamoud GA, Pureza IM, Neto NS (2020) Probabilistic evaluation of distribution power transformers reliability indices considering load transfers and mobile unit substations. Electr Power Syst Res 187:106501

21. Croitoru A (2017) Schumpeter, Joseph Alois, 1939, "Business cycles: a theoretical, historical, and statistical analysis of the capitalist process", New York and London, McGraw–Hill Book Company Inc. J Comp Res Anthropol Sociol 8(01):67–80

22. Da Silva CL, Bassi NSS (2012) Análise dos impactos ambientais no Oeste Catarinense e das tecnologias desenvolvidas pela Embrapa Suínos e Aves. Inf Gepec 16(1):128–143

23. De Andrade MP, Campos AT, Gandia RM, do Amaral Tonelli RT, da Silva Freitas LN, Ferreira JC (2019) Efficiency of batch model biodigestors in the treatment of swine manure in deep litter. Theoret Appl Eng 3(1):12–19

24. De Oliveira SVWB, Leoneti AB, Caldo GMM, De Oliveira MMB (2011) Generation of bioenergy and biofertilizer on a sustainable rural property. Biomass Bioenergy 35(7):2608–2618

25. Del Rosario Rodero M, Carvajal A, Castro V, Navia D, de Prada C, Lebrero R, Muñoz R (2019) Development of a control strategy to cope with biogas flowrate variations during photosynthetic biogas upgrading. Biomass Bioenergy 131:105414

26. Díaz-Trujillo LA, Nápoles-Rivera F (2019) Optimization of biogas supply chain in Mexico considering economic and environmental aspects. Renew Energy 139:1227–1240

27. Empresa Brasileira de Pesquisa Agropecuária (Embrapa). Available in: https://www.embrapa.br/. Accessed 14 Jun 2020

28. Freitas FF, De Souza SS, Ferreira LRA, Otto RB, Alessio FJ, De Souza SNM, Junior OA (2019) The Brazilian market of distributed biogas generation: overview, technological development and case study. Renew Sustain Energy Rev 101:146–157

29. Feyerabend PK (1970) Against method: outline of an anarchistic theory of knowledge. University of Minnesota Press, Minneapolis. Available in: https://conservancy.umn.edu/handle/11299/184649. Accessed 08 Ago 2020

30. Fu Y, Zhao J, Wang C, Peng W, Wang Q, Zhang C (2018) The virtual water flow of crops between intraregional and interregional in mainland China. Agric Water Manag 208:204–213

31. Fundação Parque Tecnológico Itaipu. Available in: https://www.iguassu.com.br/listing/fpti-funda%C3%A7%C3%A3o-parque-tecnol%C3%B3gico-itaipu/102/. Accessed 14 Jun 2020
32. Ghofrani-Isfahani P, Valverde-Pérez B, Alvarado-Morales M, Shahrokhi M, Vossoughi M, Angelidaki I (2020) Supervisory control of an anaerobic digester subject to drastic substrate changes. Chem Eng J 391:123502
33. Governo do Estado de Santa Catarina. Atividades em destaque. Available in: https://www.sc.gov.br/noticias/temas/desenvolvimento-economico/santa-catarina-ultrapassa-bahia-e-e-a-sexta-maior-economia-do-pais#:~:text=Com%20um%20crescimento%20de%204,Geografia%20e%20Estat%C3%ADstica%20(IBGE). Accessed 12 Jul 2020
34. Harari YN (2014) Sapiens: a brief history of humankind. Random House
35. Herremans IM, Reid RE (2002) Developing awareness of the sustainability concept. J Environ Educ 34(1):16–20
36. Hess DJ, Sovacool BK (2020) Sociotechnical matters: reviewing and integrating science and technology studies with energy social science. Energy Res Soc Sci 65:101462
37. Hobbes T (1998) In Tuck R, Silverthorne M (eds) Hobbes: on the citizen
38. Hoff H (2011) Understanding the nexus: background paper for the Bonn 2011 nexus conference
39. Hosseini SE (2020) An outlook on the global development of renewable and sustainable energy at the time of COVID-19. Energy Res Soc Sci 101633
40. Hyman J, Bailis R (2018) Assessment of the Cambodian national biodigester program. Energy Sustain Dev 46:11–22
41. Instituto de Tecnologia Aplicada e Inovação. Available in: https://itai.org.br/. Accessed 14 Jun 2020
42. Kajau G, Madyira DM (2019) Analysis of the Zimbabwe biodigester status. Procedia Manuf 35:561–566
43. Kneller GF (1980) A ciência como atividade humana. Zahar
44. Kotler P (2000) Administração de marketing
45. Kuhn TS (2012) The structure of scientific revolutions. University of Chicago Press
46. Lehmann S (2018) Conceptualizing the urban nexus framework for a circular economy: linking energy, water, food, and waste (EWFW) in Southeast-Asian cities. In: Urban energy transition. Elsevier, pp 371–398
47. Li C, Zhou H, Li J, Dong Z (2020) Economic dispatching strategy of distributed energy storage for deferring substation expansion in the distribution network with distributed generation and electric vehicle. J Clean Prod 253:119862
48. Liu J, Zhao D, Mao G, Cui W, Chen H, Yang H (2020) Environmental sustainability of water footprint in Mainland China. Geogr Sustain
49. Locke J (1847) An essay concerning human understanding. Kay & Troutman
50. Lonergan SM, Topel DG, Marple DN (2018) The science of animal growth and meat technology. Academic Press
51. Marques AC, Fuinhas JA, Pais DF (2018) Economic growth, sustainable development and food consumption: evidence across different income groups of countries. J Clean Prod 196:245–258
52. Martins CE, Monteiro JP (1999) Vida e Obra de Locke. Locke. Nova Cultural, São Paulo, pp 5–17
53. Maziku M, Desta S, Stapleton J (2017) Pork production in the Tanzanian livestock master plan
54. Mekonnen MM, Hoekstra AY (2010) The green, blue and grey water footprint of farm animals and animal products. Volume 2: appendices
55. Neto RDCS, Berchin II, Magtoto M, Berchin S, Xavier WG, de Andrade JBSO (2018) An integrative approach for the water-energy-food nexus in beef cattle production: a simulation of the proposed model to Brazil. J Clean Prod 204:1108–1123
56. Popper K (1992) The logic of social-sciences. Vopr Filos 10:65–75
57. Prefeitura de Itapiranga. Available in: https://itapiranga.atende.net/#!/tipo/pagina/valor/21. Accessed 14 Jun 2020
58. Rees WE (1992) Ecological footprints and appropriated carrying capacity: what urban economics leaves out. Environ Urban 4(2):121–130

59. Rousseau JJ (1772) Discours sur l'origine et les fondemens de l'inégalité parmi les hommes, vol 1. Chez Marc Michel Rey
60. Santa Catarina. Decreto nº 525, de 23 de março de 2020. Dispõe sobre novas medidas para enfrentamento da emergência de saúde pública de importância internacional decorrente do coronavírus e estabelece outras providências. Available in: https://www.coronavirus.sc.gov.br/wp-content/uploads/2020/03/DECRETO_525.pdf. Accessed 25 Jul 2020
61. Shiel C, do Paço A, Alves H (2020) Generativity, sustainable development and green consumer behaviour. J Clean Prod 245:118865
62. Simpson JR (2019) The economics of livestock systems in developing countries: farm and project level analysis. CRC Press
63. Skarzynski P, Gibson R (2008) Innovation to the core: a blueprint for transforming the way your company innovates. Harvard Business Press
64. Smith A (1759) The theory of moral sentiments. In: Millar A, Kincaid A, Bell J. AM Kelley (originally published in 1759)
65. Smith NK (2011) Immanuel Kant's critique of pure reason. Read Books Ltd
66. United Nations (UN) Available in: https://www.un.org/. Accessed 02 Aug 2020
67. Varga I, Bartha T, Soumelidis A, Katics B (2003) A concept for on-line testing of distributed safety-critical supervisory systems. IFAC Proc Vol 36(3):175–180
68. Veroneze ML, Schwantes D, Gonçalves AC Jr, Richart A, Manfrin J, da Paz Schiller A, Schuba TB (2019) Production of biogas and biofertilizer using anaerobic reactors with swine manure and glycerin doses. J Clean Prod 213:176–184
69. Wang Y, Long A, Xiang L, Deng X, Zhang P, Hai Y, Li Y (2020) The verification of Jevons' paradox of agricultural water conservation in Tianshan District of China based on water footprint. Agric Water Manag 239:106163
70. Xiong Y, Tian X, Liu S, Tang Z (2020) New patterns in China's water footprint: analysis of spatial and structural transitions from a regional perspective. J Clean Prod 245:118942
71. Zamora F (2019) Barrel aging; types of wood. In: Red wine technology. Academic Press, pp 125–147
72. Zanella C, Lopes DG, da Silva Leite AL, Nunes NA (2015) Conhecendo o campo da economia dos custos de transação: uma análise epistemológica a partir dos trabalhos de Oliver Williamson. Rev Ciênc Adm 1(2):64–77

The Social Network Analysis to Study Discourse on Water–Energy–Food Nexus

Lira Luz Benites-Lazaro, Nathália Nascimento, Alberto Urbinatti, Mateus Amaral, and Leandro Luiz Giatti

Abstract Debates on the interrelationship, interdependences, synergies, and trade-offs on water–energy–food nexus have underscored the need for new methods to explore and understand the complexity of these relevant issues, particularly methods that are capable of representing the interrelationships between sectors. In this chapter, we discuss the utility of social network analysis to analyze social actors' discourses surrounding the water–energy–food nexus sectors. Social network analysis (SNA) allows us to depict the connections that exist within the nexus sectors and those among social actors. We present a practical application of the method by analyzing textual documents regarding water, energy and food related issues from Brazilian government, media, and Non-Governmental Organizations. SNA can help present and map the degree to which coordination between actors and topics occurs in nexus systems. The essential elements of this type of analysis are centrality, and the extent to which individual nodes, which typically represent nexus topics are connected to one another in the network.

Keywords Water–energy–food nexus · WEF nexus · Biofuels · Network analysis · Nexus thinking · Bioenergy · Ethanol

L. L. Benites-Lazaro (✉) · A. Urbinatti · M. Amaral · L. L. Giatti
Department of Environmental Health, School of Public Health,
University of São Paulo, Av. Dr. Arnaldo, São Paulo 715, 03178-200, Brazil
e-mail: lbenites@usp.br

A. Urbinatti
e-mail: albertourbinatti@usp.br

M. Amaral
e-mail: mateusamaral@usp.br

L. L. Giatti
e-mail: lgiatti@usp.br

N. Nascimento
Federal University of Espirito Santo, Vitória, Brazil

1 Introduction

The provision of water, food, and energy to supply the demand of the growing global population are considered one of the greatest contemporary challenges today. The world population is expected to reach 9.7 by 2050 which, in addition to changes in consumption patterns could exacerbate human pressure on natural resources [46]. In addition to this challenge, humanity will also have to deal with the scarcity of agricultural land, the degradation of natural resources, urban problems, the impacts of climate change, and consequently, conflicts arising from disputes over natural resources [13, 36, 37].

In different parts of the world, natural ecosystems have been disturbed by agricultural expansion and infrastructure implemented to produce and distribute energy. These disturbances have triggered disastrous impacts on biodiversity and climate system [17, 28, 40]. Studies show that climate change, especially an increase in global temperatures and the occurrence of extreme weather events, is already starting to impact agricultural production worldwide, and risks compromising the productive capacity of several countries in the near future [38, 39]. Another aspect of food production is its high demand for water. Worldwide agricultural production is responsible for 70% of global water consumption, which makes it highly dependent on this resource. Scenarios produced by the Food and Agriculture Organization of the United Nation (FAO) have estimated an increase of 34% in the global area of irrigated agriculture, which should represent a 14% increase in the demand for water for irrigation [16]. In addition to being essential to human survival, water is also the main source of power generation in many countries, through hydropower [29]. Of all the energy consumed in the world, it is estimated that 30% is consumed during the food production process [16].

The nexus approach arises from an understanding that water, food and energy security are crucial to human well-being and that these three sectors are interconnected. Human well-being is understood as essential resources needed for human survival and social equity [19]. For this reason, the nexus approach has among its fundamental premises a 'responsible governance' of natural resources, integrated and participatory management of these resources, and the dialogue between different actors in favor of environmental governance [8]. Although the nexus has been gaining ground in political arenas, the creation of opportunities to integrate management, and mechanisms to involve different actors in decision-making, as well as adapting technical languages and methodologies for socio-political debate spaces, represent important challenges to concisely apply the nexus in the design of an integrated public management of the three sectors of the nexus.

However, the challenges of the nexus approach and its promise for achieving integrated decision-making and coherence in policy-making are particularly acute because sectors operate and policies are executed in areas under different institutional frameworks at different scales and through several actors with economic and political divergent interest [42, 47]. This creates the need for methods and techniques for studying and elucidating the complexity of our society, since that "approximately

80% of electronic data is in a text format." This huge amount of data, covering almost every area of human endeavor, is not only unstructured or semi-structured, but it also contains usable, useless, scientific, and trade-specific data [4]. In particular understand the discourse of actors and how to influence policies, and analyses of such debates can improve our understanding of which interests are enhanced or neglected, of what value-based arguments, for example, for and against biofuels expansion are dominant and therefore, the likely social distribution of costs and benefits in the implementation of policies [5].

Network analysis techniques can be appropriate with large datasets and become more robust when used to not only describe but also interpret [41]. The study of networks is largely interdisciplinary and has developed in many fields of human knowledge, including mathematics, physics, computer and information sciences, biology and social sciences. Network analysis has long been used as a method to examine relational data and has recently emerged as an important tool in policy analyse and governance studies [26, 34, 49]. Among the main areas in which social network analysis has been applied are to investigate of corporate power and interlocking directorships, and community structures [41]. In this chapter, we use social network analysis to examine the actor discourses related to water-energy-food (WEF) nexus relationship applied to the Brazilian context. This analysis can contribute to nexus studies due to its ability to represent the interconnectivity between different sectors such as water–energy–food nexus. The remainder of this chapter is structured as follows. In Sect. 2, we describe the methods that have been published in the literature to assess the WEF nexus. In Sect. 3, we explain the social network analysis method. Then, in Sect. 4, we present the practical application of the method to the case of discourses on the WEF nexus of biofuels nexus in Brazil. Finally, in Sect. 5, we offer the conclusions.

2 The Nexus Methods

Nexus-related research concerns two contrasting aspects of a given negotiation in relation to the interdependencies among resources. The first are the trade-offs, which can be understood as solutions from one sector that have consequences for other sectors. The second is related to synergies, that is, situations that generate positive exchanges on another sectors. According to Kurian et al. [26], although the analysis of trade-offs has the potential to reveal priorities in a governance process and to inform the norms of equity defined locally in the interventions, understanding the synergies can determine the progress in ensuring a balance and mitigating possible reverse effects in planning and environmental management. The negotiation process mainly concerns intersectorality, as it is an attempt to "break down the silos" [3]. This can be understood as a search for opening up of the specialization boxes [48]. In other words, the nexus approach considers multi-level, intersectoral, and possible transdisciplinary decision-making processes to be essential to move forward toward sustainable scenarios.

Different methods have been used in nexus-related research. One example is the Transboundary Basin Nexus Assessment (TBNA), a methodology that seeks to enable stakeholders to identify positive and negative linkages, benefits and trade-offs between relevant sectors. This framework is promoted by the United Nations Economic Commission for Europe [45]. These linkages are identified and mapped in a qualitative way through the participation of experts and officials. This enhances the possibility of establishing a basis for quantifying those linkages. In short, "the methodology further provides for governance assessments aimed at increasing the understanding of how a coherent integration of sectors might be achieved" [45, p. 15]. The main pillars of this approach are: participatory processes, knowledge mobilization, sound scientific analysis; capacity building; collective effort; and benefits and opportunities.

Concerning quantitative analysis, Karnib [24] shows the so-called Q-Nexus Model can suggest practical and mathematically-based quantitative WFE nexus assessments. This model can serve as a "platform to quantify, plan, simulate and optimize water, energy and food as an interconnected system of resources that directly and indirectly affects one another" [23, p. 89]. It also helps to evaluate WEF planning scenarios and policy options. With respect to textual analysis, a good example is a study conducted by Benites-Lazaro et al. [4], which analyzed documents from public and private agents by combining topic modeling method with discourse analysis. The authors confirmed the usefulness of these techniques to capture main topics and extract dominant discourses among the actors involved in the documents analyzed. The topic modeling method can be applied to the study of WEF nexus for discovering and exploring the thematic structures as well as to navigate through time to uncover how discussions related to nexus sector have evolved over the years [4].

The literature review conducted by Albrecht et al. [1] showed that the use of specific methods to carry out nexus assessment is unusual. There are more studies that emphasizes the quantitative approaches. This highlights that WEF nexus is a field of study that is still in development; therefore, it is understandable that there is disagreement when it comes to the methods used. Endo et al. [15] analized review articles about nexus and identified five main groups of methods: comprehensive reviews; targeted reviews; synthesis; articles assessing the interlinkages, trade-offs, and/or synergies among resources geographically; and case studies. Urbinatti et al. [47] is an example of a targeted review as it seeks to systematically review the concept of governance of nexus from the analysis of social networks. Although it is a widely used tool for analysis of individuals and organizations, there are recent efforts to understand the potential of this methodology also for systematic reviews [11, 30]. The bibliographic conceptual study carried out by Urbinatti et al. [47] about nexus governance helped to stress the trans-scalar nature of the issue, the need for more critical definitions, and the perspective that the nexus remains challenging in terms of being a scientific debate navigating opportunities and constraints to influence decision-making processes.

Previous studies shown the prioritization of quantitative methodologies over qualitative ones, observed both globally [1] and in the global south [12]. This priorization contributes to the development of a less holistic perception of the social, environmental and economic dimensions of the provoked trade-offs by allocating resources between the three domains of the nexus. Therefore, it becomes relevant to consider how social sciences and interdisciplinary approaches can contribute toward obtaining more promising results that can be applied in decision-making for integrated resource management [1]. These epistemological challenges, externalized by the nexus approach, are often justified by a lack of sufficient information [2] or specific methodologies and tools for analysis, and can be justified by the sectorialization of the knowledge and management models applied to the management of resources over the years. Therefore, we understand that by testing methodologies from different contexts and territorial sections it can contribute toward the optimization of research applied to the nexus.

3   Social Network Analysis

Social network analysis (SNA) is the study of social structures, specifically the interactions and networks between a group of actors (individuals or organizations), represented by nodes, which are tied by some way by common interests, such as financial exchanges, friendship, dislike, knowledge or prestige [10, 22]. Social networks operate on many levels, from family relationships and diseases spreading to the level of corporate strategies, social movements or even nations.

SNA is a way to re-incorporate context and bridge the gap between the micro and the macro, the cells constituting an animal, the individuals constituting groups, or the actors constituting a political system. Thus, the method allows researchers to retain the traditional units of recording but simultaneously broadens the perspective by including information about the relationships across units. The additional structural information gained from this method allows researchers to address existing research questions with new tools and to approach them from different theoretical perspective [9].

SNA mixes sociology and mathematics in an attempt to simplify sociological concept and explain macro-sociologicaly issues from a micro-sociology perspective. In SNA analysis, social relations are conceived in terms of vertices and edges, and mathematical graphs (called "graphs"), whereby actors are represented by vertices and connections are represented by either edges or arcs (with directions) [10, 41]. Actors can either be individuals or groups of individuals, for example, companies, communities, social organizations, countries, or cities. Connections are relationships among actors, for example, power, partnerships, family kinship, e-mail contact, common religious beliefs, or rivalry. Moreover, connections may have one or more weights. The result produced are structures based on graphs or sociograms, sometimes very complex and difficult to interpret, and has achieved significant results thanks to the joint of innumerable researchers from various fields, in particular from the field of computer science.

Network analysis focuses on calculating all the basic properties of a given network, such as the diameter of the graph and the geodesic distances (shortest path lengths), as well as more advanced structural statistics, such as the centrality of the vertices in the network and the prestige indexes (Centrality of proximity, Centrality of Interest, Prestige Proximity, etc.), and groupings and triads (grouping coefficient, triad census) [9]. Historically, SNA was one of the first analytical tools in using graph theory [41]. It emerged due to the need to make the social sciences somewhat more formal, which gave rise to sociometry. One of its strands used statistics to study populations (macro-level), another used graph theory to model relationships between people (micro-level), following the idea of the genealogical trees of anthropology [9, 41].

There are at least four ways to approaching the study of social networks, either by the general characteristics that it presents, by the position of the actors, by the groups that conform it, or by the visualization of them. First, it is possible to identify that there are many types of networks that have not yet been classified, but there are already very frequent phenomena in them. An example of such social networks is the so-called six degrees of separation [25]. This phenomenon is also known as *small world* [33] and occurs in networks with special connectivity whereby the half distance between two actors is very small in comparison to the size (number of actors) of the network. In other networks, such as quotes from scientific articles, and the internet among, the distribution of degrees follows a power law or similar. These networks are called scale-free networks because in a subgraph of this type of network, the degrees are still likely to be distributed as power law. Scale-free networks are interesting because they repeat in many other cases, such as in the distribution of tickets (Effect Matheus) [32]. Phenomena with these properties continue to be observed time and again.

Second, the concept of the *locational position* of an actor in a network corresponds to the access it has to the rest of the network. In principle two actors are known to occupy the same place in a given network if they share the same neighbors (i.e., structural equivalence, which is a local version of edge isomorphism). Measures of centrality have typically been used as proxies for influence and power, and have enabled research to be carried out on brokerage relationship [41]. A person located in the center of a star is assumed to be structurally more central than any other person in any other position in any other similar sized network [20]. There are four measures of centrality that are widely used in network analysis: degree centrality, betweenness, closeness, and eigenvector centrality. Centrality measures attempt to quantify how central each person or topic is inside a given social network. To that end, these measures usually examine both the ties attached to an actor as well as the geodesic distances (shortest path lengths) to other actors [22].

Third, the detection of communities, groups, ghettos (exclusive groups), etc. that form the network, are of great interest in the study of social networks. It is complicated because it is not easy to define a group. The definition becomes easy when there is a formal structure involved, or when there is a defined group and "a group" of adherents that are said to be part of the group (e.g., Brazil and Brazilians). However, this becomes complex, obscure, and even esoteric when it comes to informal structures. A group of friends represent a lot of people, in which all or most are friends with one another, but members of this group also have friends in common from outside the group, therefore it can be difficult to defines who belong to the group and who does not. There are many techniques to detect groups, and many algorithms that obey different ideals of how to detect a group, and these techniques are very opposed to measures the centrality [50].

A traditional way is to reduce group detection to a classification or clustering. Within these techniques are the traditional k-means, genetic algorithms, modularity analysis (number of links between groups is small, high within groups), among others. In addition, these techniques are parametrizable (number of classes in k-means, minimum modularity, etc.), which allows analyzing the quality of the classification. Here, hierarchical trees can be used to decide when the ranking is good. Another traditional way to detect a group is to apply graph theory, for example, by using different colors to classify groups in graphs. In this case, it is also possible to see the problem as one of structural equivalence transformed into another of regular equivalence: "in a group of friends, the friends share the same friends" (this defines an iterative algorithm). More recent forms include the use of measures of centrality: centrality of proximity, "in one group, all actors are close;" centrality of intermediation, "one group is a network more or less isolated from the rest;" and democratic criteria, which consist of counterbalancing two or more different group detection criteria.

Fourth, the visualization of social networks also serves as a method of approach that allows discovering properties, but has less theoretical weight in the analysis. Observing complex networks is a big challenge. In general, much information as posible is visually presented so that it is understandable. There are many ways to look at the data, and each can illustrate different analytic properties: such as centrality, communities, key actors, etc. When it comes to detecting communities, there are a large number of algorithms that allow their visualization, each one obeys a different idea and purpose. But often the goal is to gain an immediate presentation of the network and to have an efficient algorithm. In this study, we used SonecTv software.

4 Practical Application of Network Analysis in Nexus Issues—The Case of Brazilian Biofuels

Brazil is a leader in the export of various agricultural products such as soybeans, corn, and coffee [31]. However, the country also leads the list of countries with the largest losses of forests in the world in the last years. In 2019 alone, Brazil was responsible for more than a third of all global deforestation. The expansion of the agricultural frontier is the main driver of deforestation in the country and has compromised springs, bodies of water, and the ability of forests to generate rain. The deterioration of water resources directly affects the electric energy sector in Brazil, given that 65% of its energy is generated by hydroelectric plants [14]. Due to the exporter of agricultural commodities conditions, the nexus approach is needed to explore to the case of Brazil, for example, connections between demands for resources and the supply of exports. These commodity exportation practices directed toward supplying external markets, generate pressure on local resources, for example on the public supply of water and energy, and on the provision of fresh and healthy food for large populations in the more urbanized areas.

With water scarcity as the central focus of nexus relations [21], water can be considered to have an irrefutable complexity, which implies that there are no simplified solutions and no trade-offs to dialogue with nexus components. In Brazil, despite being a country that is very rich in terms of water resources, contradictorily, there is a great irregularity in the distribution and demand for water resources in the immense national territory. Highly complex regions that share an interdependence on resources, such as the São Paulo Macrometropolis, can attest to the seriousness of the trade-off relationship and threats to sustainability. In the vast territory of the state of São Paulo, for example, which is home to more than 30 million inhabitants in its 180 municipalities, it has been faced with critical scenarios in the public supply of water, energy, and land availability, especially considering the regional role of the bioethanol industry [5, 44].

The nexus approach makes it possible to recognize global relationships of scarcity which in turn are interdependent. Thus, the growing global demand for water, energy, and food matches situations that transcend geographical scales. In local contexts, for example, the insufficiency associated with the nexus inevitably concerns extrapolates territorial limits, given the dynamic way water, energy and food are interrelated in globalization, for example in the case of the import and export of agricultural commodities.

The Brazilian government and sugarcane ethanol companies have constructed national narratives about ethanol production. These narratives invoke a connection between the environmental benefits of ethanol and its influence on social well-being with respect to the development of modern technologies for its production. However, the use of agro-energy has raised fundamental questions pertaining to the impact of ethanol production, especially with regard to the demand for land and water needed for the production process and the required guarantee of a continuous supply of food crops.

In order to understand how the nexus approach has been applied in Brazil, we used the SNA technique. The data sources were text documents addressing issues surrounding biofuels, food security, climate change and water crisis, and were obtained from Brazilian newspapers labeled "MEDIA," reports from non-governmental organizations, labeled "NGO," government websites and documents, labeled "GOV," search with keywords in Portuguese language (biocombustíveis, etanol, segurança-alimentar, mudança climatica, crise hídrica, escacez de água). We used web scraping technique to freely access webpage such as those of the government, and NGO. To scrape a website, we developed scripts using R, an open-source programming language and software environment for statistical computing and graphics. We employed the rvest package to search for HTML code that generates a webpage. To search for websites in JavaScript, we used the PhantomJS package. These packages provide an immediate and easy-to-use solution to search for content generated by free-access web pages. Access to newspapers with copyrights, was through two Brazilian universities; the data (2007–2016) was provided by the Observatory on Public Policies for Agriculture at the Rural Federal University of Rio de

Janeiro, and then updated with data published up to December 2019, provided by the University of São Paulo. All this data constitutes our corpus for the analysis. In this study, we used T-Lab to lemmatization, text treatment and adjacent matrix creation to network analysis, and SocneTv to network visualization. There are several other open-source software packages that can be used, for example, Iramuteq or Unicet.

This type of analysis of documents from different social actors has been important to unveil conditions that corroborate certain reductionism regarding the inherent complexity of productive chains. For example, although bioenergy production alone is really important as a global guideline for reducing GHG emissions, this segment may be developing in a way that ignores water scarcity or inherent trade-offs and competition with land demand for food production. In this sense, this study explored hypotheses regarding the distinct roles and discourses of different social actors (such as media, government, and companies). It can contribute to understand of the need to identify insufficient discourses, which in turn correspond to incomplete (or even competing) agendas in the face of the breadth and interdependence that must be recognized in the pursuit of sustainable development.

Figure 1 shows the network of keywords applied to the corpus using the T-Lab's tool sequence and network analysis, which takes into account the positions of the various lexical units relative to one another, and permits to represent and explore any text as a network. From this Figure, it is possible to observe links between words such as water resources, biofuels, food, ethanol, government, politics, among others.

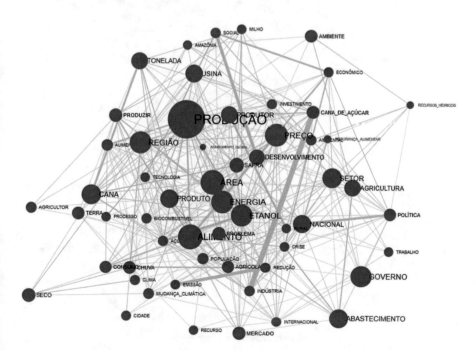

Fig. 1 Network analysis. *Source* Authors

With this general result applied to the corpus, ego networks were carried out for the keywords "energy" and "water," as shown in Fig. 2. Ego network enables a description of the relationship and role of the "ego" (keyword) in its social network. The network is composed of one keyword centring the graph (the ego), all users linked to this ego (called "alters"), and all relations between the alters (keywords). Among the objective of using this method was to identify the most important keywords linked to energy, water, food, climate change and health (Fig. 3). Then, we carry out the ego network to these themes and for each actor (Fig. 4).

These results (Figs. 1 and 2) demonstrate that water, energy, and food are interdependent and not easily disentangled. Thus, any strategies or decisions made by only

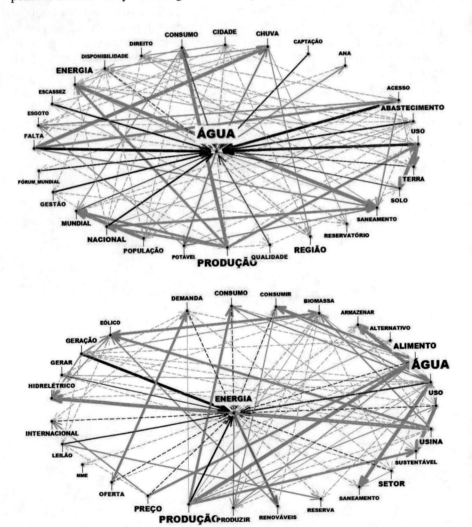

Fig. 2 Ego network of keywords water and energy. *Source* Authors

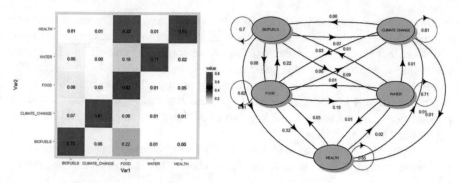

Fig. 3 Thematic network. *Source* Authors

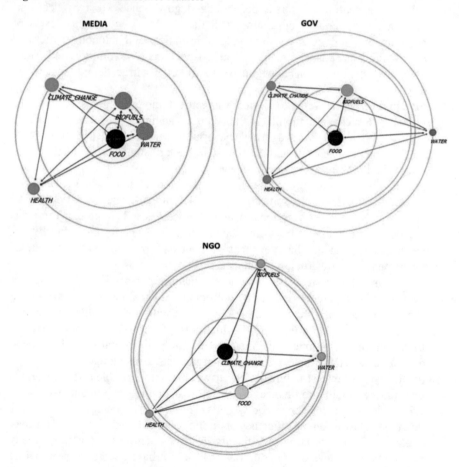

Fig. 4 Degree centrality of the thematic network by actors. *Source* Authors

focusing on one of the nexus sectors without considering the interconnections and interdependences among them could have serious consequences. One of the main problems that the nexus approach seeks to overcome is the tradition of carrying out sectorial work and decision making in silos, focusing on only one sector (for example, energy/biofuel) without considering how any related policies could negatively or positively affect other sectors (e.g., water/land use) [5]. This issue of interdependence between sectors is a key aspect of the nexus approach and has been categorized as trade-offs.

According to Kurian and Ardakanian [27] these trade-offs are reflected in the following five choices that are often made when it comes to determining public policies:

a) Prioritizing centralized versus decentralized governance, whereby decision-making power remains concentrated in the hands of both higher levels of government and donors. As a result, local initiatives and autonomy suffer and any planning and adaptive management prospects become compromised. Thus, political decentralization has begun to gain importance in academic and political discussions.

b) Prioritizing public versus private management models. In response to the growing disenchantment with centralized management, the world, particularly Latin America, underwent a phase where public services were privatized. With this, regulatory agencies, and the primacy of establishing self-regulation were established in different sectors of the economy.

c) Prioritizing infrastructure rather than services, particularly when it comes to service parameters such as affordability, reliability, and quality. Services have been overlooked by conventional planning processes and structures. There has been a prioritization of large investments in infrastructure, including hydro-electric dams, water and wastewater treatment, and irrigation, at the expense of investments in infrastructure maintenance.

d) Prioritizing short-term versus long-term planning perspectives. Planning and decision-making are carried out by politicians and business leaders who have short-term thoughts; they are more concerned with their political career, re-election, or the accounting period, which has a fiscal year-end close by date. In recent years, the methods employed to compute capital costs have been questioned and there have been forceful arguments to take longer term views of the life cycle of infrastructure projects to ascertain the possible revenue streams that may be possible to finance infrastructure operations and maintenance.

e) Prioritizing efficiency versus equity goals. The emphasis on building infrastructure tends to focus on the efficiency of public services and the system. This may explain why despite large investments in infrastructure, there are still many communities in developing countries that do not have access to basic public services. Thus, interest in community-based natural resource management has led to an interest in issues such as equity in the distribution of benefits generated by these natural resources.

Using the concept of trade-offs it is possible to examine issues of justice and its relationship to environmental management by analyzing land-use, land tenure, water use, access to markets and the equity issues related to management of the benefits and costs of environmental management, and the land-water competitions in the biofuel production as this chapter try to demonstrate. The trade-offs within the complex interactions of the nexus also tell us about the consequences for human well-being, health, poverty and inequality. Food production systems, for example, rely heavily on energy for fertilizers, water for irrigation, and the functions of nature to recycle nutrients and pollinate crops [6, 18]. Altering the availability or demand of one sector can have repercussions on the others. Thus, recognizing the relationships between the elements of the nexus poses challenges that require considering beyond the three elements when evaluating the consequences and planning investments, policies, and actions. Such cross-cutting challenges include, for example, a strong push toward the use of biofuels, which could affect the availability of land and water for other purposes, especially for food production, extensive areas of irrigated agricultural land that use water and are needed for food production and energy generation, particularly hydroelectric power [7].

Figure 3 shows the thematic network of water, food, health, biofuels and climate change, and highlights the interdependencies and connections between them. Nexus approach is based on an understanding of the synergies and regulated negotiation of fair trade-offs between competing uses of water, energy and food. However, other components or sectors can also be taken into account. As Fig. 3 shows, health can be an important component within the nexus, as well as labor or land that is independent when studying the WEF nexus of biofuels. In this study, the health theme is related to the expansion of agriculture, the use of pesticides or agrochemicals, and the contamination of water, the latter of which is occurring a large and constant scale. The study reveals that agrochemicals are easily spread in the environment, and thereby, reach the atmosphere and contaminate the areas surroundings where they are applied thanks to their potential to volatilize from the soil, water and plant surface. This process can consequently impact the health of the sorrounding population.

Figure 4 shows the degree centrality of the thematic network by actors. In terms of graphs, the degree centrality of an actor is calculated by its number of neighbors. For example, in a model of a social network of friends, the centrality of each actor would consist of the number of friends that each individual actor possesses, and the individual who has the most friends or acquaintance (highest popularity) would have the highest degree centrality index [20]. This study examined 3 different types of actors: The media, the government and NGOs. With respect to the MEDIA, three themes dominated discussion on the nexus approach: Food, Biofuels and Water. Moreover these themes demonstrated to be well connected one another. These three themes are the basis of the nexus sector. It is possible that the advancement of discussions on this issues and the importance of inserting it in the public debate on food production, energy and water resources protection has been dominating the communication channels in the Brazilian media.

For the government, "food" was also found to be a central theme. This could be related by the importance of agricultural sector in Brazil. Another theme that came up a lot was "biofuels," a theme that is closely related to food. In Brazil, sugarcane is the basis for ethanol production. It is also one of the most important agricultural crops in the southeastern part of the country. More recently, the revision of zoning sugarcane plantation in fragile ecological biomes such as Pantanal and the Amazon was revised, thus directly linking the theme of ethanol to the problems of agricultural expansion and deforestation in the country.

With respect to NGOs, "climate change" was found to be the central theme. NGOs have a discourse that is more aligned with international debates, where concerns about water, food, and energy are debated in light of the risks of climate change. The "Climate change" theme was followed by "food", thus confirming the importance of food for the three groups.

Surprisingly, despite the crucial importance of water for the other two sectors, the theme "water" does not appear as a protagonist in any of the three groups, and only appears as a main theme in the MEDIA thematic network. The most worrying is that for government, i.e., the actor among the three groups with a great deal of decision-making power, water is the least popular theme, after climate change and health. This demonstrates the frailty of water governance in Brazil compared to the food and energy production sectors. Moreover, it demonstrates the lack of government measures to adapt to a future projected to be impacted by climate change.

In 2014, for example, a rare anomaly in the high atmospheric pressure system resulted in an abrupt decrease in precipitation in southeastern Brazil and generated a crisis in the water supply system and compromising the irrigation system across the state of São Paulo, which is the leader in sugarcane production in the country [35]. This event generated a domino effect, bringing to the fore problems in water management in southeastern Brazil. Even so, our analyses show that the theme "water" is not yet close to the leading role in the Brazilian government discussions. An early study of Benites-Lazaro et al. [5] found that in Brazil the water and land abundance narratives, along with narratives surrounding the country's tropical climate, have been appropriated to grow and expand the bioenergy and agribusiness sectors. Additionally, the dominant perception among Brazilians that water is an abundant resource, as is often supported by members of the political and economic elite, has created both a lack of awareness and inaction when it comes to water conservation and management, and has contributed to poor water governance.

The results of this study reveal network analysis to be a useful tool to explore the connection between themes and keywords. However, despite this technique being useful to depict relationships, connections, and links and provide quantitative measures of thematic and keyword relationships, it does not reveal anything about the context in which the words are used in a given text. As previous studies [4, 43] have shown, there is a need to incorporate other methods, such as discourse analysis, to qualitatively analyze the discourses of the actors analyzed.

5 Conclusion

This study found SNA to be a good and easy way to both understand the distribution of important topics related to the nexus approach and the gaps in that distribution. Moreover, this study reveals SNA to be a useful tool to assess differences between groups, gaps in themes, and centrality of specific themes. A curious aspect of these results is the role that "food" theme played in all the groups considered in the analysis and the low centrality of the theme "water", especially with respect to the actor "government". The results reveal how certain themes and sectors are overlooked. Thus, there is a need to take into account the synergies and trade-offs of the nexus sectors when it comes to planning and managing these resources, so as not to cause stress by prioritizing one sector to the detriment of another. A limitation of the method is its inability to provide details of contexts and discourses, in which these words, sentence, fragment or paragraph were inserted. Such information would facilitate the discussion and and lead to a better understanding of why certain themes dominate over others. In any case, the method can be very useful when the researcher is already aware of the context and discourses of each group and uses the SNA to assess differences between groups, gaps in themes, and centrality of specific themes.

Brazil is an interesting place to apply the nexus approach. The country is the top exporter of various agricultural products to various countries in the world and is considered one of the main global agricultural powers. Moreover, in Brazil the population is increasingly urban, and the country has been witnessing high rates of deforestation and natural resource degradation, which, combined with the impacts of climate change, demand urgent integrated planning among the agricultural, energy and water sectors if it is to remain an agribusiness power and guarantee the food, water and energy security of its population. Future studies based on the results of this study could focus on taking an in-depth analysis of the nexus themes. The method used in this study could also be applied to cases other than biofuels. The network method could also be applied to quantitative data, which would contribute toward an understanding of the relationship between the nexus sectors.

Acknowledgements The authors acknowledge financial support received from São Paulo Research Foundation (FAPESP), Process: 2015/03804-9. The first author also acknowledges the Grant: 2017/17796-3 São Paulo Research Foundation (FAPESP), Brazil.

References

1. Albrecht TR, Crootof A, Scott CA (2018) The water-energy-food nexus: a systematic review of methods for nexus assessment. Environ Res Lett 13:043002
2. Arthur M, Liu G, Hao Y, Zhang L, Liang S, Asamoah EF, Lombardi GV (2019) Urban food-energy-water nexus indicators: a review. Resour Conserv Recycl 151:104481
3. Artioli F, Acuto M, McArthur J (2017) The water-energy-food nexus: an integration agenda and implications for urban governance. Polit Geogr 61:215–223

4. Benites-Lazaro LL, Giatti L, Giarolla A (2018) Topic modeling method for analyzing social actor discourses on climate change, energy and food security. Energy Res Soc Sci 45:318–330
5. Benites-Lazaro LL, Giatti LL, Sousa Junior WC, Giarolla A (2020) Land-water-food nexus of biofuels: discourse and policy debates in Brazil. Environ Dev 33:100491
6. Bervoets J, Eveillé F, Thulstrup A (2018) Strengthening the Water-Food-Energy-Ecosystems (WFEE) nexus
7. Bizikova L, Roy D, Swanson D, Venema HD, McCandless M (2013) The water-energy-food security nexus: towards a practical planning and decision-support framework for landscape investment and risk management. International Institute for Sustainable Development
8. Bonn Conference (2011) Bonn2011 conference: the water, energy and food security nexus— solutions for a green economy. Bonn
9. Borgatti SP, Everett MG, Johnson J (2013) Analyzing social networks. Sage
10. Borgatti SP, Mehra A, Brass DJ, Labianca G (2009) Network analysis in the social sciences. Science (80-)
11. Cowhitt T, Butler T, Wilson E (2020) Using social network analysis to complete literature reviews: a new systematic approach for independent researchers to detect and interpret prominent research programs within large collections of relevant literature. Int J Soc Res Methodol 23:483–496
12. Dalla Fontana M, de Araújo Moreira F, Di Giulio GM, Malheiros TF (2020) The water-energy-food nexus research in the Brazilian context: what are we missing? Environ Sci Policy 112:172–180
13. Diehl P (2018) Environmental conflict: an anthology. Routledge. https://doi.org/10.4324/978 0429500794
14. EPE (2018) Anuário Estatístico de Energia Elétrica 2018 no ano base de 2017. Empresa de Pesquisa Energética (EPE)
15. Endo A, Yamada M, Miyashita Y, Sugimoto R, Ishii A, Nishijima J, Fujii M, Kato T, Hamamoto H, Kimura M, Kumazawa T, Qi J (2020) Dynamics of water–energy–food nexus methodology, methods, and tools. Curr Opin Environ Sci Health 13:46–60
16. FAO (2017) The future of food and agriculture: trends and challenges. Roma
17. Fehlenberg V, Baumann M, Gasparri NI, Piquer-Rodriguez M, Gavier-Pizarro G, Kuemmerle T (2017) The role of soybean production as an underlying driver of deforestation in the South American Chaco. Glob Environ Change 45:24–34
18. Flammini A, Puri M, Pluschke L, Dubois O (2014) Walking the nexus talk: assessing the water-energy-food nexus in the context of the sustainable energy for all initiative. Fao
19. Food and Agriculture Organization of the United Nations (2014) The water-energy-food nexus. https://doi.org/10.1016/j.envsci.2016.11.006
20. Freeman LC (1978) Centrality in social networks conceptual clarification. Soc Netw 79:215–239
21. Hoff H (2011) Understanding the nexus. Background paper for the Bonn2011 nexus conference. Stockholm Environment Institute, pp 1–52
22. Kalamaras D (2020) Social network visualizer: SocNetV manual [WWW document]. URL https://socnetv.org/docs/index.html. Accessed 8.10.20
23. Karnib A (2017) Water, energy and food nexus: the Q-nexus model. Eur Water 60:89–97
24. Karnib A (2018) Bridging science and policy in water-energy-food nexus: using the Q-nexus model for informing policy making. Water Resour Manag 32:4895–4909
25. Kleinfeld J (2002) Could it be a big world after all? The six degrees of separation myth. Society, Apr 12, 5–2
26. Kurian M, Portney KE, Rappold G, Hannibal B, Gebrechorkos SH (2018) Governance of water-energy-food nexus: a social network analysis approach to understanding agency behaviour. In: Managing water, soil and waste resources to achieve sustainable development goals: monitoring and implementation of integrated resources management
27. Kurian M, Ardakanian R (2015) Institutional arrangements and governance structures that advance the nexus approach to management of environmental resources. In: Dresden Nexus Conference 2015. pp 57–68

28. Laurance WF, Sayer J, Cassman KG (2014) Agricultural expansion and its impacts on tropical nature. Trends Ecol Evol. https://doi.org/10.1016/j.tree.2013.12.001

29. Lee M, Keller AA, Chiang PC, Den W, Wang H, Hou CH, Wu J, Wang X, Yan J (2017) Water-energy nexus for urban water systems: a comparative review on energy intensity and environmental impacts in relation to global water risks. Appl Energy. https://doi.org/10.1016/j.apenergy.2017.08.002

30. Leppink J, Pérez-Fuster P (2019) Social networks as an approach to systematic review. Health Prof Educ 5:218–224

31. MDIC—Ministério da indústria comércio exterior e Serviços (2020) Balança Comercial [WWW document]. Balança Comer

32. Merton RK (1968) The matthew effect in science. Science (80-) 159:56–63

33. Milgram S (1967) Small world problem. Psychol Today 2:60–67

34. Mischen P, Jackson S (2008) Connecting the dots: applying complexity theory, knowledge management and social network analysis to policy implementation. Public Adm Q 32:314–338

35. Nascimento N, Ometto JP (2019) Governança ambiental do Nexus-Água, Alimento e Energia: proposta metodológica para análise baseada em Redes bayesianas aplicada na Macrometrópole Paulista. In: I Fórum de Governança Ambiental Da Macrometrópole Paulista. São Paulo, pp 145–149

36. Ouda S, Zohry AE-H, Noreldin T, Ouda S, Zohry AE-H (2020) Water scarcity leads to food insecurity. In: Deficit irrigation. Springer International Publishing, pp 1–13. https://doi.org/10.1007/978-3-030-35586-9_1

37. Porkka M, Gerten D, Schaphoff S, Siebert S, Kummu M (2016) Causes and trends of water scarcity in food production. Environ Res Lett 11:015001. https://doi.org/10.1088/1748-9326/11/1/015001

38. Ray DK, West PC, Clark M, Gerber JS, Prishchepov AV, Chatterjee S (2019) Climate change has likely already affected global food production. PLoS ONE 14:e0217148. https://doi.org/10.1371/journal.pone.0217148

39. Renard D, Tilman D (2019) National food production stabilized by crop diversity. Nature 571:257–260. https://doi.org/10.1038/s41586-019-1316-y

40. Sauer S (2018) Soy expansion into the agricultural frontiers of the Brazilian Amazon: the agribusiness economy and its social and environmental conflicts. Land Use Policy 79:326–338. https://doi.org/10.1016/j.landusepol.2018.08.030

41. Scott J (2011) Social network analysis: developments, advances, and prospects. Soc Netw Anal Min 1:21–26

42. Sharma P, Kumar SN (2020) The global governance of water, energy, and food nexus: allocation and access for competing demands. Int Environ Agreements Polit Law Econ 20:377–391

43. Törnberg A, Törnberg P (2016) Muslims in social media discourse: combining topic modeling and critical discourse analysis. Discourse Context Media 13:132–142

44. Travassos LRFC, Zioni SM, Torres PHC, de Souza Fernandes B, Araujo GM (2020) Heterogeneity and spatial fragmentation in the Sao Paulo macrometropolis: the production of borders and holes. Ambient Soc 23:e01801

45. United Nations (2018) Methodology for assessing the water-food-energy-ecosystem nexus in transboundary basins and experiences from its application

46. United Nations, Department of Economic and Social Affairs, Population Division (2017) World population prospects: the 2017 revision, methodology of the United Nations population estimates and projections. New York

47. Urbinatti AM, Benites-Lazaro LL, de Carvalho CM, Giatti LL (2020a) The conceptual basis of water-energy-food nexus governance: systematic literature review using network and discourse analysis. J Integr Environ Sci 1–23

48. Urbinatti AM, Dalla Fontana M, Stirling A, Giatti LL (2020b) 'Opening up' the governance of water-energy-food nexus: towards a science-policy-society interface based on hybridity and humility. Sci Total Environ 744:140945

49. Varone F, Ingold K, Jourdain C (2017) Studying policy advocacy through social network analysis. Eur Polit Sci 16:322–336

50. Wellman B, Leighton B (1979) Networks, neighborhoods, and communities: approaches to the study of the community question. Urban Aff Rev 14:363–390

The German Capability Index—An Operationalization of Sen's Capability Approach

Holger Schlör⊙, Wiltrud Fischer⊙, and Sandra Venghaus⊙

Abstract When reflecting on the global call for sustainable development (Gutteres in Davos speech 2019, January 24. World Economic Forum, Davos, 2019 [29]), a question posed by Amartya Sen actuates further thought—namely that for who has to be sustained (Anand and Sen in World Dev 28:2029–2049, 2000a [4])? This question was answered by himself, proposing "it is not so much that humanity is trying to sustain the natural world but rather that humanity is trying to sustain itself (Sen in J Hum Dev Capabil 14:6–20, 2013 [66])". Against this background, in our research we addressed this question of who has to be sustained using Sen's capability approach for the case study of Germany. For this purpose, the German household survey (Einkommens- und Verbrauchsstichprobe (EVS)) of the German Federal Statistical Office is used to analyse five social household groups (all households, single households, single parents, couples without children, couples with children) according to their income (nine income classes) (Federal Statistical Office Germany in Wirtschaftsrechnungen. Einkommens- und Verbrauchsstichprobe Einnahmen und Ausgaben privater Haushalte 2018, 2020a [19]; Federal Statistical Office Germany in Wirtschaftsrechnungen. Einkommens- und Verbrauchsstichprobe Konsumausgaben privater Haushalte 2018, 2020b [20]). Based on this survey, the question of who needs to be sustained will be analysed as follows:

1. First, the "worldwide reach (Rosa in Resonanz: Eine Soziologie der Weltbeziehung. Suhrkamp, Berlin, 2019 [62])" of the consumption patterns of the German household groups will be measured by their ecological and water footprints.

H. Schlör (✉) · S. Venghaus
Institute of Energy and Climate Research, Systems Analysis, IEK-STE, Forschungszentrum Jülich, 52425 Jülich, Germany
e-mail: h.schloer@fz-juelich.de

S. Venghaus
e-mail: s.venghaus@fz-juelich.de

W. Fischer
Department for Political Sciences and Sociology, University of Bonn, Lennéstraße 25-27, 53113 Bonn, Germany
e-mail: s5wifisc@uni-bonn.de

© The Author(s), under exclusive license to Springer Nature Singapore Pte Ltd. 2021   145
S. S. Muthu (ed.), *The Water–Energy–Food Nexus*,
Environmental Footprints and Eco-design of Products and Processes,
https://doi.org/10.1007/978-981-16-0239-9_6

2. Second, an operationalization is derived for Amartya Sen's sustainable development definition using the food–energy–water nexus as the core of sustainable development (United Nations (UN Water) in Water, food and energy. United Nations, 2020 [73]), and to reveal both

    a. potential contradictions between the FEW sectors and other sectors of the German society which can negatively affect sustainable development in Germany, and
    b. the degree of inequality in functionings and capabilities among German households.

3. Finally, those household groups will be identified which "have to be sustained" the most according to Sen's definition.

In the presented model, based on ul Haq (Fukuda-Parr in Fem Econ 9:301–317, 2003 [24]; Haq in Human development in a changing world. UNDP, New York, 1992 [30]; Haq in Reflections on human development. Oxford University Press, Oxford, 1995 [31]) and Sen (Anand and Sen in Human development index: methodology and measurement. Occasional papers. UNDP, New York, 1994 [3]; Sen in Ökonomie für den Menschen [Economy for the people], 2nd edn. dtv, Munch, 2003 [59]), the German capability index (GCI) is used to reveal both the capabilities and functionings of German society and the underlying justice structure of the German society. The capabilities (realization opportunities) in this context describe people's opportunities of using their functionings (abilities) to achieve a place in society (Venkatapuram in Health justice. Polity Press, Cambridge, 2011 [81]) given their personal capabilities, while simultaneously maintaining a sustainable use of the food–energy–water resources. The aim of the German capability index (GCI) is to make the capabilities of the various German households comparable and establish the GCI as an index of revealed capabilities.

Keywords Food–energy–water nexus · Capability approach · Germany · Sustainable development

1 Introduction

Secretary-General of the United Nations Antonio Gutteres expressed his concerns about the state of the world [27] in Davos 2019, however, a question posed by Amartya Sen actuates further thought—namely that for who has to be sustained [4]? This question was answered by himself, proposing "it is not so much that humanity is trying to sustain the natural world but rather that humanity is trying to sustain itself [60]", because "the precariousness of nature is our peril, our fragility [60]".

This, however, would inevitably imply accounting for inter- and intragenerational justice and thus requires the setting of ecological restrictions on the use and the management of the food–energy–water-nexus (FEW-nexus) resources in order to reduce (as will be discussed in further detail below) the current as well as future "worldwide reach" [55] of consumption and production patterns.

Hence, for humanity to sustain itself, especially the core of sustainable development—namely the food–energy–water nexus [70]—must first and foremost be transformed to become sustainable. This core is currently affected by global challenges as described in the IPCC Reports [32–34] and reflected in the UN Sustainable Development Goals [69].

With respect to human development, these challenges endanger the vision of the UN with respect to "the rule of law, justice, equality and … of equal opportunity permitting the full realization of human potential and contributing to shared prosperity [69]" and thus the chances of a decent life around the globe.

Based on the ideas of Kant [35], development means providing people with an environment which enables them to live a long, healthy and creative life [3], which enlarge people's choices and opportunities [3, 71]. Hence, an essential differentiation must be made between the means which can be used for a certain development and the ultimate aim of this development: the welfare of the people [60, 71].

Against this background, sustainable development can be defined in line with Sen's capability approach as a "development that prompts the capabilities of present people without compromising capabilities of future generations [60]". Intragenerational social justice corresponds to "equality of capabilities [45]", which are not restricted by the consumption patterns of others.

So far, Sen's idea of capabilities has been applied among others in the context of energy justice and sustainable development by various scientists: Altman uses Sen's capabilities approach in order to analyse socioeconomic development and the well-being of individuals. It is of interest for Altman to highlight the correlation of individual capabilities and the imagination of a decent life with the concept of economic welfare [1]. Bartiaux present a theoretical and empirical framework for a nexus among energy justice and capabilities [10]. This statistical operationalization of Sen's and Nussbaum's concept of capabilities examines all households of Belgium; nevertheless, it is transferable to other countries as well. Also, Biggs et al. present an integrated assessment approach for the nexus of water, energy, food and a sustainable decent life [11]. Bartiaux et al. examine energy policy-related inequalities of different social classes in Belgium and Portugal against the background of various justice theories [9]. Hillerbrand enlarges the scope of the capability approach by including reflections on sustainability and its criticisms [31]. This way the synergies between the environment and individuals' well-being are included. Sovacool and Dworkin propose concepts of energy justice as a base for a conceptual and decision-making tool to better integrate procedural and distributive justice concerns [67]. Martínez-Guido et al. introduce a newly developed optimization approach including the water–energy–food nexus in order to improve the Human Development Index [44]. Thereby they take into account environmental and social sustainability criteria.

Nevertheless, it can overall be stated that the specific combination of Sen's capability approach with the concept of the FEW-nexus is not yet explicitly found in the corresponding research disciplines. The present study thus represents a first attempt to operationalize these two concepts in order to close this research gap.

Thus, in the following the capability approach will be presented. Its objective is to leave the traditional welfare economics behind, which interpret the standard of living only in terms of consumed goods [64]. In the first step, the conceptual background of Sen's capability approach will be discussed and its content-related connection to the concept of worldwide reach of Hartmut Rosa will be elucidated. In the second part, the two-step approach is presented to define the current status of the German FEW-Nexus sectors and the impact of the German households on food, energy and water resources. In a third step, the German capability index (GCI) is presented, which enables the comparison of direct and indirect effects of the German households' consumption patterns in one index. Finally, some concluding remarks and an outlook of further research recommendations will be made.

2 Conceptual Background—The Capability Approach

2.1 Capabilities, Functionings and Their Worldwide Reach

Capabilities (realization opportunities) describe a person's opportunities of using her or his functionings (abilities) to reach her or his place in society given the respective personal skills [64]. They describe the possible paths of life available to a person given her or his abilities. As such, capabilities are thus enablers of the "worldwide reach expansion" of the modern times, defined by Hartmut Rosa as the categorical imperative of the modern age [54]: Act at any time so that your world in reach becomes larger [55]. This is done by the multiplication of goods, contacts and options which increase with growing functionings. This new idea of modernity brings "more world within reach", to make more world understandable and consumable. Accordingly, the consumption decisions of the households affect societies and communities around the world [56].

Hartmut Rosa believes that the modern age is characterized by the central endeavour of making the world more accessible in all social spheres and areas of life [54]. In this sense, the human quest is a systematically increase of the range of what is recognizable, controllable, attainable and available (in education, medicine, media, politics, sports, etc.) based on the earned income [56]. The worldwide reach increases with rising income. The available income and technology enable an increase in reach of the households and are a symbol of the freedom of choice.

The higher the income the higher the worldwide reach of the consumption patterns and the impact on the functionings and capabilities of people outside the own community and society. Thus, we revert to the capability approach to define measures for quantifying the worldwide reach of the German households.

2.2 The Capabilities—A Vector of Abilities

Capabilities can therefore be described as a vector of an individual's abilities, which are, in turn, a function of basic social conditions (democracy, freedom) [14, 58] and their worldwide reach: $\left(v_{\text{capabilities}}^{\text{individual}} \overrightarrow{(\text{functionings, worldwide reach})} \right)$.

Similar to a person who has a lot of money and can purchase many goods, a person with many capabilities can choose between different life concepts [14], as the following example explains [14, 52]: A wealthy person who is fasting may have the same functional performance as a poor person with respect to nutrition. However, whereas the poor person may have to starve, the wealthy person has the capability to choose whether to eat or fast. Hence, the wealthy person has a greater worldwide reach than the poor person, who has no choice [53, 64]. In the same way, a person who is only equipped with the most basic functionings, e.g.

$\left(v_{\text{capabilities}}^{\text{poor person}} \overrightarrow{(\text{functionings (food, clothes, basic education, low worldwide reach))}} \right)$

has very little opportunity of becoming, e.g. an architect or a surgeon. Her or his capabilities are limited because of the personal functionings [53]. The individual capability budget is small, whereas the capability budget of, e.g., the architect is significantly higher expressed in a sophisticated capability vector:

$\left(v_{\text{capabilities}}^{\text{architect}} \overrightarrow{(\text{functionings (food, clothes, higher education, high worldwide reach))}} \right)$

The capability budget describes the different possible combinations of functionings (abilities) a person can achieve [42]. The budget contains the means which allow a person to live a certain life [38], or the paths of life which remain closed to him because of his limited capability budget [39]. The capability budget of a person therefore describes the possibilities of a person to live the life she or he wishes [50, 62].

These functionings enable people to achieve an income and enable consumption decisions to purchase the goods that go along with a certain worldwide reach. The functionings therefore refer to the possibility of fulfilling individual needs. Moreover, a distinction is made between people's basic needs (food, clothes and basic education) and further abilities such as playing an instrument or speaking a foreign language.

Sen uses his capability approach to specify his justice approach, defining social justice as "equality of capabilities [46]" and assume basic rights to be essential for a good life [59]. The opportunities of people are expressed in their capability budget [50], which defines people's welfare and "the person's freedom to lead one type of life or another [58]", based on institutional conditions. Amartya Sen and Martha Nussbaum try to capture people's well-being in their capability approach by considering not only people's income but also non-monetary aspects (values, capabilities) which influence people's welfare and their quality of life [50]. Hence, the composition of the capability budget is a question of institutional justice [61, 63, 64], because as Jean-Baptiste Henri Lacordaire stated: "Entre le fort et le faible, entre le riche et le pauvre, entre le maître et le serviteur, c'est la liberté qui opprime et la loi qui affranchit [40]".[1] Sen's welfare approach defines welfare from two sides [38]: The outcome people can achieve based on their functionings or based on the opportunities, they have according to their capability budget. Hence, welfare is a function of outcome and opportunities.

In a next step, based on the model of Sen and the methods set up by Mahbub ul Haq in the conceptual framework of the human development report and its Human Development Index (HDI), we derive a method to reveal and to measure the capabilities and functionings of the German society including the worldwide reach. We use our revealed functioning approach to capture the following functionings based on the Human Development Index: welfare, nourishment, decent and healthy life, mobility, social cohesion and education to define social justice as equality of capabilities.

3 Measuring the Worldwide Reach of German Households on the FEW-Nexus

The proposed two-step approach first measures the current state of the FEW-nexus sectors and the impact of the German households on food, energy and water. The approach follows the objective to consider both the direct effects of the consumption of food, energy and water measured by German household survey data, but also the indirect effects caused by the German consumption patterns measured by the ecological footprint and the water footprint. Both measures belong to the environmental footprint indices [47, 48].

Secondly, it reveals the functionings and capabilities of the German households through the German capability index (GCI).

3.1 German Household Survey (EVS)

For our analysis, we use the latest German household survey data "Einkommens- und Verbrauchsstichprobe 2018" (EVS) of the German Statistical Office [19, 20] to make functionings observable, measurable and comparable. The household survey EVS records the income and expenditures of all German households [18]. The EVS 2018 is the latest survey of the Federal Statistical Office of Germany and delivers important data for the assessment of the income situation, the standard of living and the expenditure behaviour of the whole population and the different households [24]. The EVS database therefore provides information about German economic life and consumer behaviour [18]. The EVS methodology is based on the Eurostat recommendations for "Household Budget Surveys in the EU [16]".

---

[1]Between the strong and the weak, between the rich and the poor, between the lord and the slave, it is freedom which oppresses and the law which sets free.

The EVS database characterizes a household [18] as an individual with own income who economizes for herself/himself as well as a group of related or personally linked people who belong together according to their income as well as according to their consumption. The latter must live together and dispose together of one or several incomes or income shares as well as supply a joint household [18]. Households are analysed according to their social characteristics and their net income [18]. The net income of the households is calculated by subtracting income tax, church tax, solidarity tax and social insurance from the gross income [18]. The disposable income of the chosen social groups is identified by adding to the net income the earnings from good sales as well as other income modes. Not included in the disposable income are earnings from the liquidation and conversion of monetary and material property as well as from borrowing [18]. The savings of the households cover the expenses for the development of financial and tangible assets and repayment of credits [18].

For this analysis, we use the data of 40.6 million households in Germany covering the following household groups: all households (40.6 million), single women (10.6 million), single men (6.5), single parents (1.2 million), couples without children (11.7 million) and couples with children (5.8 million) with further differentiation between the following income groups: under 900, 900–1300, 1300–1500, 1500–2000, 2000–2600, 2600–3600, 3600–5000, 5000–18,000 € [19, 20]. The group of other households[2] (4.9 million) will not be analysed. We use the data on income and expenditures of these households to define their material functionings: earnings, savings, food, beverages, tobacco, clothing and shoes, living, household tools and furniture, health, transport, communications, entertainment, leisure time, education, restaurants and services [18]. The data "living without energy" contains the hot, cold and waste water costs of the households, but it is not possible to further differentiate between these respective water costs. Hence, the following FEW-nexus expenditure costs are the minimum expenses of the households for the FEW-nexus goods.

Table 1 shows that FEW-nexus expenditures account for more than 20% of the expenditures of the average German household. In the lowest income group, the households spend on average 30% of their consumption expenditures for the FEW-nexus resources. With growing income, the nexus share decreases to 24% in the highest income group. The importance of the FEW-nexus expenditures for the households decreases with increasing income. The FEW-nexus commodities are relative inferior[3] goods.

Table 2 shows the expenditure structure of the single woman households. The single woman households spend on average 352 € for FEW-nexus products. In the lowest income group, the households spend only 251 € per month for FEW-nexus commodities. This value increases with rising income to 514 € in the highest income group. On average 21.4% of the household consumption expenditures are allocated to the FEW-nexus goods. This percentage share has its highest value with 28.4%

---

[2]Other households are those household which have further household members (e.g. parents-in-law, children of full age) [18].

[3]A detailed explanation of inferior, normal and superior goods can be found by Krugman and Wells [37] and Wied-Nebbeling [84].

Table 1 All households

All households according to their net income, 2018 from … to …, in €

	All households	Under 900	900–1300	1300–1500	1500–2000	2000–2600	2600–3600	3600–5000	5000–18,000
Extrapolated households, in 1000	40,683	2006	3413	1816	4803	5475	7250	6895	9024
Food, beverages	360	170	193	220	249	295	354	434	540
Clothing and shoes	122	29	42	55	68	87	109	148	229
Living without energy[b]	762	336	409	459	526	618	755	902	1164
Energy living	146	63	78	93.5	118	167	270	306	372
Interior equipment, household issues	137	22	40	53	67	91	126	165	269
Health care	50	7	16	18	28	41	58	82	117
Mobility without fuel	308	21	53	59	102	123	186	306	554
Fuel	71	41	48	60	70	115	150	170	208
FEW-nexus expenditures[a]	577	274	319	373.5	437	577	774	910	1120
Communication	71	37	43	50	54	61	70	83	99
Leisure and culture	304	65	99	131	166	218	285	372	557
Education	28	8	6	7	11	14	21	33	63
Accommodation and catering	168	36	49	60	80	113	146	205	334
Other commodities and services	111	32	46	60	67	85	107	129	194

(continued)

Table 1 (continued)

All households according to their net income, 2018 from … to …, in €

	All households	Under 900	900–1300	1300–1500	1500–2000	2000–2600	2600–3600	3600–5000	5000–18,000
Total consumption	2704	901	1152	1352	1625	2016	2551	3252	4657
Expendable income	3726	747	1125	1427	1786	2331	3125	4327	7739
FEW-nexus expenditures in % of total consumption	21.3	30.4	27.7	27.6	26.9	28.6	30.3	28.0	24.0

Source German Statistical Office (2020), and own calculations. [a]Food, beverages, energy living, fuel. [b]Contains the hot, cold and waste water costs; IEK-STE 2020

Table 2 Single women

Single woman households according to their net income, 2018 from … to …, in €

	All households	Under 900	900–1 300	1300–1500	1500–2000	2000–2600	2600–3600	3600–5000	5000–18,000
Extrapolated households, in 1000	10,595	1041	2058	938	2353	1927	1495	522	260
Food, beverages	209	159	179	196	210	227	236	259	275
Clothing and shoes	78	35	46	58	74	93	114	138	190
***Living without Energy*[b]**	560	328	405	453	532	623	766	892	1047
Energy living	108	80	92	100	107	115	127	144	158
Interior equipment, household issues	78	22	42	53	71	93	118	164	224
Health care	72	26	33	42	55	71	109	181	393
Mobility without fuel	128	41	59	79	112	150	211	312	329
Fuel	45	12	23	33	45	55	73	81	81
FEW-nexus expenditures[a]	362	251	294	329	362	397	436	484	514
Communication	50	35	41	45	50	54	60	64	74
Leisure and culture	189	64	97	149	174	223	295	362	474
Education	8	7	5	5	7	8	10	12	29
Accommodation and catering	84	36	41	59	74	103	130	170	216

(continued)

Table 2 (continued)

Single woman households according to their net income, 2018 from … to …, in €

	All households	Under 900	900–1 300	1300–1500	1500–2000	2000–2600	2600–3600	3600–5000	5000–18,000
Other commodities and services	84	38	49	69	76	92	120	144	301
Total consumption	1692	884	1112	1342	1585	1907	2370	2921	3791
Expendable income	2077	754	1117	1432	1771	2306	3068	4200	8420
FEW-nexus expenditures in % of total consumption	21.4	28.4	26.4	24.5	22.8	20.8	18.4	16.6	13.6

Source German Statistical Office (2020), and own calculations. [a]Food, beverages, energy living, fuel. [b]Contains the hot, cold and waste water costs; IEK-STE 2020

in the lowest income group, the share decreases to 13.6% in the highest income group. The relativity of the inferior good characteristic of the FEW-nexus goods is increasing.

The single man households (Table 3) show a similar consumption structure as the single woman households. Single men spend on average 380 €—22% of total consumption—for FEW-nexus commodities. In the lowest income group (under 900 €), this spending decreases to only 270 € for FEW-nexus commodities, whereas this value increases with rising income and reaches its highest value with 510 € in the highest income group, while the economic significance decreases with rising income. The FEW-nexus share decreases from 32% in the lowest income group to 15% in the highest income group.

The meaning of the FEW-nexus for the consumption decisions of the households decreases with rising income. The relativity of the inferior character is not so distinct as in the group of single women.

For the single parent's households, a slightly different picture is revealed (Table 4). The average single parent's household uses 24.2% (535 €) of its consumption budget for FEW-nexus expenditures. In the lowest income group, the households spend 364 € for FEW-nexus good; i.e. these households use on average 28.3% of their consumption budget for FEW-nexus goods.

This share decreases also for the single parents to 19.9% in the highest income group but the absolute expenses of the households increase with rising income from 364 € in the lowest income group (900–1300) to 877 €. The highest income group spends more than twice the amount of the lowest income group for FEW-nexus goods. The FEW-nexus goods change their character for this household group with rising income. For the households with a monthly net income between 2000 and 2600 € the FEW-nexus goods are relative superior commodities, the meaning of these goods increases with growing income but for the households with higher income the meaning diminishes and changes into a relative inferior good.

In the following, we will analyse, which picture is created by the analysis of the couple households.

The couples without children spend on average 693 € on FEW-nexus related goods (Table 5). This accounts to only 21.7% of the consumption budget for these commodities, which is lower than in the case of the single parent households. The households of the lowest income group spend nearly 28% of their consumption budget for FEW-nexus products, but, atypically in this social group, the saturation point of the highest share spent on these goods is reached in the second lowest income group and after this peak the share decreases slightly to 24% in the highest income group. However, as in the previous social groups the absolute amounts spent for the FEW-nexus increase continuously up to the highest income group (1096 €).

For the couples without children, the FEW-nexus commodities are slightly relative inferior goods.

The couples with children present a different picture in contrast to the previous social groups. The average households spend 22.6% of their consumption budget for the FEW-nexus goods. These households spend on average 867 € for these commodities. This spending reaches its saturation point in relative terms in the income group

Table 3 Single men

Single man households according to their net income, 2018 from … to …, in €

	All households	Under 900	900–1300	1300–1500	1500–2000	2000–2600	2600–3600	3600–5000	5000–18,000
Extrapolated households, in 1000	6472	909	1023	460	1036	1047	1036	595	368
Food, beverages	217	175	193	204	222	229	234	242	261
Clothing and shoes	49	21	28	40	44	54	65	77	95
Living without energy[b]	548	326	398	444	479	576	683	820	946
Energy living	102	82	87	92	101	106	112	130	125
Interior equipment, household issues	70	18	32	51	59	78	104	117	161
Health care	71	15	24	30	43	53	83	160	352
Mobility without fuel	184	35	92	98	133	200	293	369	411
Fuel	61	13	26	43	62	77	87	97	124
FEW-nexus expenditures[a]	380	270	306	339	385	412	433	469	510
Communication	53	37	44	52	54	57	61	66	71
Leisure and culture	184	57	102	115	172	198	257	317	397
Education	8	7	7	6	7	9	9	9	11
Accommodation and catering	123	34	63	71	91	139	166	240	305

(continued)

Table 3 (continued)

Single man households according to their net income, 2018 from … to …, in €

	All households	Under 900	900–1300	1300–1500	1500–2000	2000–2600	2600–3600	3600–5000	5000–18,000
Other commodities and services	79	32	53	81	81	101	120	148	171
Total consumption	1729	843	1127	1290	1512	1834	2237	2735	3406
Expendable income	2355	759	1116	1423	1796	2316	3115	4281	7337
FEW-nexus expenditures in % of total consumption	22.0	32.0	27.2	26.3	25.5	22.5	19.4	17.1	15.0

Source German Statistical Office (2020), and own calculations. [a]Food, beverages, energy living, fuel. [b]Contains the hot, cold and waste water costs; IEK-STE 2020

Table 4 Single parents

Single parents households according to their net income, 2018 from … to …, in €

	All households	Under 900	900–1300	1300–1500	1500–2000	2000–2600	2600–3600	3600–5000	5000–18,000
Extrapolated households, in 1000	1234	Null	76	114	294	291	261	135	24
Food, beverages	333	Null	232	257	305	339	352	422	321
Clothing and shoes	49	Null	28	40	44	54	65	77	95
***Living without energy*[b]**	639	Null	412	439	517	591	728	924	1158
Energy living	141	Null	90	103.5	132	233.5	282	282	350
Interior equipment, household issues	99	Null	36	47	55	99	112	195	378
Health care	60	Null	18	22	28	41	57	119	283
Mobility without fuel	165	Null	130	44	128	125	148	352	484
Fuel	61	Null	42	65.5	83	180.5	166	170	206
FEW-nexus expenditures[a]	535	Null	364	426	520	753	800	874	877
Communication	71	Null	50	57	62	74	74	92	133
Leisure and culture	240	Null	93	100	150	237	309	395	373
Education	41	Null	0	0	28	38	47	91	91
Accommodation and catering	109	Null	34	38	64	101	144	197	122

(continued)

Table 4 (continued)

Single parents households according to their net income, 2018 from ... to ..., in €

	All households	Under 900	900–1300	1300–1500	1500–2000	2000–2600	2600–3600	3600–5000	5000–18,000
Other commodities and services	103	Null	56	56	71	98	133	153	268
Total consumption	2208	Null	1288	1326	1692	2100	2489	3408	4411
Expendable income	2601	Null	1217	1419	1795	2313	3063	4224	6670
FEW-nexus expenditures in % of total consumption	24.2	Null	28.3	32.1	30.7	35.9	32.1	25.6	19.9

Source German Statistical Office (2020), and own calculations. [a]Food, beverages, energy living, fuel. [b]Contains the hot, cold and waste water costs; Null = not enough data available; IEK-STE 2020

Table 5 Couples without children

Couples without children households according to their net income, 2018 from … to …, in €

	All households	Under 900	900–1300	1300–1500	1500–2000	2000–2600	2600–3600	3600–5000	5000–18,000
Extrapolated households, in 1000	11,722	Null	194	214	755	1475	2920	2953	3180
Food, beverages	415	Null	299	315	339	363	400	427	474
Clothing and shoes	134	Null	53	54	70	84	105	140	206
Living without energy[b]	866	Null	Null	524	563	641	785	910	1122
Energy living	167	Null	97	112	111	151	264	300	354
Interior equipment, household issues	174	Null	68	65	73	101	136	172	281
Health care	173	Null	46	35	53	84	108	163	329
Mobility without fuel	331	Null	154	62	107	153	243	352	552
Fuel	111	Null	44	53	71	115	235	248	268
FEW-nexus expenditures[a]	693	Null	440	480	521	629	899	975	1096
Communication	75	Null	51	53	57	62	69	79	91
Leisure and culture	380	Null	111	113	155	236	305	399	586
Education	11	Null	8	10	13	9	9	11	15
Accommodation and catering	221	Null	54	56	86	121	158	230	370

(continued)

Table 5 (continued)

Couples without children households according to their net income, 2018 from … to …, in €

	All households	Under 900	900–1300	1300–1500	1500–2000	2000–2600	2600–3600	3600–5000	5000–18,000
Other commodities and services	131	Null	67	56	68	90	106	131	199
Total consumption	3189	Null	1583	1505	1781	2174	2686	3311	4560
Expendable income	4389	Null	1187	1418	1806	2358	3140	4318	7593
FEW-nexus expenditures in % of total consumption	21.7	Null	27.8	31.9	29.3	28.9	33.5	29.4	24.0

Source German Statistical Office (2020), and own calculations. [a]Food, beverages, energy living, fuel. [b]Contains the hot, cold and waste water costs; Null = not enough data available; IEK-STE 2020

of 2600–3600 €, whereas the expenses in absolute terms grow continuously until the highest income group. This social group spends on average 1217 € for FEW-nexus related commodities (Table 6).

For the couples with children, the meaning of the FEW-nexus commodities also changes with rising income. Until an income of 1500–2000 €, the goods have a relative superior character, and then the character changes to a relative inferior good, similar to the single parents.

In the following, we will analyse the indirect effects of the consumption patterns of the German households on the FEW-nexus sectors not only in Germany but also on a global level. To do so, we analyse in the following the effect of the household consumption expenses and its impact on the ecological footprint and the water footprint.

3.2 Ecological Footprint

The central starting point for the development of the ecological footprint was the recognition that people are part of nature and they depend on it and its goods (energy, wood, materials and water) and use nature as a repository for their waste [81, 82]. Water and energy are two life-support services of nature without which life would not be possible [6, 66]. Hence, ecological sustainability inevitably requires living within the boundaries of nature [7, 8]. The ecological footprint serves as a measure for this [48].

Wackernagel describes the ecological sustainability approach in the development of the ecological footprint as follows: "Not living within our ecological means will lead to the destruction of humanity's only home. Having insufficient natural resources, not living decently and equitably will cause conflict and degrade our social fabric. Thus, the ecological footprint as an ecological sustainability tool determines whether "people's quality of life improves over time" [79]". The improvement of the human quality of life should take place within the ecological possibilities of the ecosystem. Wackernagel's approach calculates the impact that humans have on the earth and its resources using the ecological footprint (EF) [79].

Different organizations have also developed methods to calculate ecological footprints for the earth, countries, regions and communities [74].[4] In the following, the procedure developed by Wackernagel and Rees in 1996 and continued by the Global

---

[4]https://www.myfootprint.org/.
   https://www.ecologicalfootprint.org/.
   https://www.redefiningprogress.org/.
   https://www.agenda21berlin.de/fussabdruck/vorstellung.html.
   https://www.deutschebp.de/sectiongenericarticle.do?categoryId=9008474&contentId=701 5566.
   https://www.ecobusinesslinks.com/ecological_footprint_calculator.htm.
   https://www.ew.govt.nz/enviroinfo/indicators/community/sustainability/ecofoot/report.htm.
   https://www.bestfootforward.com/foot.html.

Table 6 Couples with children

Couples with children households according to their net income, 2018 from … to …, in €

	All households	Under 900	900–1300	1300–1500	1500–2000	2000–2600	2600–3600	3600–5000	5000–18,000
Extrapolated households, in 1000	5790	Null	86	85	136	365	867	1561	2824
Food, beverages	536	Null	325	357	392	449	474	518	585
Clothing and shoes	202	Null	66	93	88	108	133	178	255
Living without energy[b]	996	Null	Null	Null	528	610	721	913	1203
Energy living	178	Null	111	125	140	123	286	306	352
Interior equipment, household issues	212	Null	108	78	65	86	132	175	283
Health care	117	Null	18	15	30	42	51	79	174
Mobility without fuel	435	Null	224	145	82	226	267	323	596
Fuel	153	Null	41	50	109	109	246	264	280
FEW-nexus expenditures[a]	867	Null	477	532	641	681	1006	1088	1217
Communication	95	Null	78	68	73	86	91	92	100
Leisure and culture	420	Null	105	119	109	184	242	365	555
Education	104	Null	Null	16	48	50	71	87	133
Accommodation and catering	234	Null	44	31	74	88	131	178	325

(continued)

Table 6 (continued)

Couples with children households according to their net income, 2018 from … to …, in €

	All households	Under 900	900–1300	1300–1500	1500–2000	2000–2600	2600–3600	3600–5000	5000–18,000
Other commodities and services	151	Null	79	67	63	89	108	133	187
Total consumption	3832	Null	1702	1498	1779	2275	2714	3367	4759
Expendable income	5602	Null	1180	1426	1808	2388	3189	4391	7668
FEW-nexus expenditures in % of total consumption	22.6	Null	28.0	35.5	36.0	29.9	37.1	32.3	25.6

Source German Statistical Office (2020), and own calculations. [a]Food, beverages, energy living, fuel. [b]Contains the hot, cold and waste water costs; Null = not enough data available; IEK-STE 2020

Footprint Network is presented [81], as it also serves as the basis for the WWF Living planet report [78]. For the Global Footprint Network, the ecological footprint accounting system is "a method for calculating society's use of nature's assets [12]", as the food, energy and water resources.

Wackernagel defines the ecological footprint as follows: "It compares humanity's ecological footprint (the demand our consumption places on the biosphere) with biocapacity (the biosphere's ability to meet this demand), providing a kind of bank statement for the planet [12]". The ecological footprint thus adds up the various resources consumed by a given population and expresses them in units of productive land needed to provide these resources and to absorb its waste. The ecological footprint defines sustainability as a measure for the use of nature by humans: "The ecological footprint is a measure of how much biologically productive land and water area an individual, a city, a country, a region, or humanity requires to produce the resources it consumes and to absorb the waste it generates, using prevailing technology and resource management schemes. This land and water area can be anywhere in the world [76]". Hence, "the global ecological footprint is [a] consequence of the increasing human demand for food, fibre, energy and water [75]" and hence, a measure for the FEW-nexus.

The biocapacity measures the ability of the earth to regenerate itself and the ecological footprint measures now the available biocapacity as well as the global demand for this capacity. "The common measurement unit is global hectares: biologically productive hectares with world average productivity. Thanks to this common measurement unit, countries, regions, cities, individuals and products can be compared across the world and over time [78]".

The ecological footprint is based on seven sub-indices [80]: Cropland Index, Grazing-Land Index, Fishing Index, Forest Index and the Indices for the capture of waste (CO_2, nuclear) and the Build-up Index [36, 77]. The fossil fuel footprint is calculated on the basis of the area required to sequester the CO_2 emissions from fossil fuels, minus the CO_2 emissions absorbed by the oceans [77]. Hence, the ecological footprint measures the global impact of the current production and consumption patterns on the food, energy and water sectors worldwide. The ecological footprint put the focus on the current ecological realities on the stressed ecosystems [12].

The Global Footprint network calculated that in 2019 ca. 12.2 billion hectares of biologically productive land and water were available, so that every person has at his/her disposal 1.6 global hectares [25]. This area also accommodates "the wild species that compete for the same biological material and spaces as humans [26]". The ecological footprint of a country or region summarizes the biological productive area on earth, which is needed to provide the resources for its consumption patterns under the current economic conditions [26].

3.3 Ecological Footprint Germany

Based on these considerations, the Global Footprint Network detected for Germany that the German population causes an annual ecological footprint of 4.5 ha [25, 43]. This implies that Germany is living beyond its means and needs nearly the resources of three Earths to provide the annually consumed resources. But the German households do not contribute equally to this footprint, as Fig. 1 shows [43].

Figure 1 shows that the average ecological footprint of the average household is 8000 m^2 per month, so that the average footprint per person is 4000 m^2, but only 1333 m^2 of biocapacity per month are globally available. Figure 1 also shows that the footprint increases continuously with rising income and no social group is below the available biocapacity. The footprint increases for the group of all households from 2666 m^2 for the households in the lowest income group to 13,778 m^2 in the highest income group. To measure the differences between the households, we measure the spread between the ecological footprint of the highest and lowest income group of the various social groups.

Hence, the footprint of the highest income is 5.2 times higher than in the lowest income group and 1.7 times higher than of the average household. For the single households, the ecological footprint increases from 2615 and 2494 m^2 to 11,216 and 10,077 m^2 in the highest income group. The spread between the highest and the lowest income group is 4.0 and 4.3 for women and men, respectively. For the single parents, the ecological footprint increases also from 3811 m^2 in the lowest income group to 13,050 m^2 which results in a spread of 3.4 between the lowest and highest income group. The couples without children cause on average an ecological footprint of 9435 m^2, which increases from 4683 m^2 in the lowest income group to 13,491 m^2 in the highest income group. The couples with children produce on average an ecological footprint of 11,337 m^2, and the footprint reaches its highest peak at 14,080 m^2 in the highest income group. The spread between the lowest and

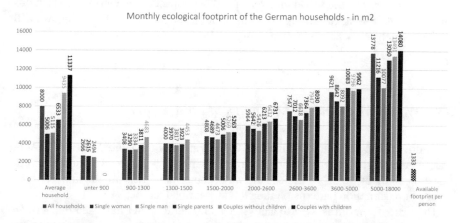

Fig. 1 Ecological footprint Germany. *Source* Own estimation based on Lin et al. [43]

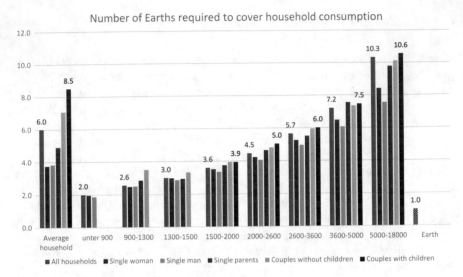

Fig. 2 Number of Earths *Source* Own estimation based on Lin et al. [43]

highest income is smaller (2.9, 2.7), because for the lowest income group no data is available. In the following, it will be analysed how many earths are needed to cover the consumption of the households differentiated according to their income, their social position and their size which varies between the social groups.

Figure 2 shows that the average household needs six earths to cover its consumption needs, whereas the average couple household with children needs more than eight earths to cover its consumption needs. In the lowest income group, the households (single men and women, and single parents) need two earths for their daily consumption. This value increases with rising income to more than ten earths in the highest income group for the couples with children.

These data reveal the worldwide reach and the impact of the German households on the global FEW-nexus through its consumption patterns. In the following, we will analyse which impact the German consumption patterns have on the global water system.

3.4 Water Footprint—The Centre of the FEW-Nexus

According to the Food and Agriculture Organization of the United Nations (FAO), water is the most important basic resource on earth [17]. To account for the exceptional meaning of water, we choose the water footprint to measure the impact of the German consumption patterns on the global water budget.

The water footprint is used to analyse the amount of water that used by German households not only directly through their personal water consumption, but also through the commodities they consume [23]: "Everything we use, wear, buy, sell and eat takes water to make [83]".

Hence, the water footprint measures the water consumption along the value chain of the production of industrial and agricultural goods as well as of the consumption of the various commodities. The water footprint is an indicator for the use of the resource water. The indicator makes the hidden water trade at the expense of water-poor countries more transparent [23].

For Germany, it is the German Environmental Agency (UBA) that calculates the water footprint. Germany's total water footprint is 117 billion cubic metres (cbm) of water per year [23], which corresponds to a daily consumption of nearly 4000 L per inhabitant. Furthermore, more than half of the water for the products and goods we consume does not originate from Germany. It represents instead the external water footprint of Germany through imported goods. The UBA detected that most of the water used in Germany is attributed to agricultural goods from Brazil, the Ivory Coast and France [23].

In the following chapter, it will thus be analysed how much the various social groups according to their social status and income contribute to the German water footprint.

3.4.1 The Water Footprint of the German Households

Figure 3 shows the monthly water footprint of German households depending on their social position and their household income.

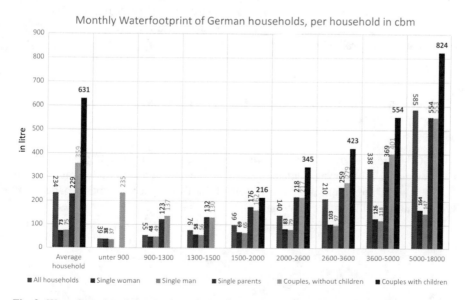

Fig. 3 Water footprint of German households. *Source* Own estimation based on UBA [23]

Figure 3 shows the estimation of the various levels of water consumption according to German household groups. The estimation is based on the calculations of the UBA [23]. The average German household consumes 234 cbm of water per month, whereas single woman and single man households consume only 73 and 75 cbm a month, respectively. Single parent households, in contrast, consume on average 229 cbm, households of couples without children 359 cbm, and couples with children 631 cbm of direct and indirect water. Figure 3 also shows that the water footprint increases continuously with rising income. In the group of all households, the households of the highest income group use on average 2.5 times more water than the average household of that social group. This spread is also valid for the group of single parents (2.4). In the case of single women and single men, these households use 2.0 and 2.2 times more direct and indirect water than the average household of this social group. This spread decreases for the couple households without children and for the couple households with children to 1.5 and 1.3, respectively. We thus conclude that the couple households share similar needs which are more independent of the income than is the case for other households. Analysing the relationship between direct and indirect water consumption between the highest and the lowest income group (the lowest income group with data available), we receive a different picture. The spread between the highest and the lowest income group in the consumption of direct and indirect water is 15 for the social group of all households. The highest income group consumes 15 times more water than the households in the lowest income group. This spread decreases to values around 4 for the single households and decreases further for the couple households to 3.8 and 2.4. Reverting to the idea of the worldwide reaches as discussed above, the worldwide water reach of German household's increases significantly with growing income.

So far, we have developed the data basis to estimate the capability index of the FEW-nexus in Germany, considering both direct and indirect effects of the German consumption and production patterns.

4 German Capability Index (GCI)—A Measure of the Worldwide Reach

The GCI aggregates the water footprint, the ecological footprint and the data of the German household survey data into one index. The Capability Index enables the comparison of direct and indirect impacts of the German consumption patterns of German households on the FEW-nexus.

4.1 Methodological Background

The methodological background of the presented approach is the Human Development Index (HDI) of the United Nations Development Programme (UNDP). The HDI of the UNDP is published regularly in the Human Development report (HDR) [71]. The concept of the HDI is based methodologically on Sen's capability approach [68, 71].

The UNDP selected three essential conditions (i.e. functionings, namely health, education and access to natural resources (energy, water, etc.)) for human development, which have a decisive influence on people's capabilities:

- "to lead a long and healthy life,
- to acquire knowledge and
- to have access to natural resources needed for a decent standard of living [49]".

These UNDP functionings must be part of the capability budget and thus define the life people can live. If these three conditions are not fulfilled, people's options are reduced or eliminated. In addition, according to UNDP, for human development further aspects are of central significance: political, economic and social freedom, the opportunity to have possessions, be productive and creative and to have guaranteed human rights [71]. A decent life cannot be reduced to the earned income as the sum of human life [71]. UNDP argues in the tradition of Aristotle that "wealth is evidently not the good we are seeking; for it is merely useful and for the sake of something else (Aristotle 350 BC)". That "something else" is the opportunity "to realize the full potential of every human life, not just of a few, nor of most, but of all lives in every corner of the world—now and in the future [13]".

Therefore, we can summarize that UNDP and Sen do not focus their analysis on the goods people produce and consume, but Aristotle's "something else", which enables them to realize their potentials to live the life they want [2–5, 50, 62]. The life they are capable of living is determined by their own set of functionings and values. In this welfare concept, "the lives the [household] lead ... is of intrinsic importance, not the commodities or income that they happen to possess. Income, commodities ('basic' or otherwise), and wealth do of course have instrumental importance but they do not constitute a direct measure of the living standard itself [3]". The living standard is defined by the choices people have to live the life they want and to enlarge thereby their worldwide reach.

We allocated the income, the expenditure and their FEW-nexus connectivity of the German households to the functionings derived from the UNDP, as Table 7 shows.

The welfare basis in our approach is defined by the earnings and savings of the household, and a decent life is defined by expenditures for clothing, living and furniture. Nutrition is covered by food, beverages and tobacco expenditures, whereas mobility is covered by transport costs and education by expenditures for culture and education. Social participation is described by expenditure for communications, leisure, services, restaurants and hotels. The FEW-nexus activities of German households are embedded in the direct functionings nutrition and mobility and the indirect

Table 7 Functionings, expenditures, ecological and water footprint and the FEW-nexus

Functionings and expenditures of German households

Functionings	Expenditures	FEW-Nexus
Welfare basis	Earnings, savings	
Decent life	Clothing, living, furniture	
Nutrition	Food, beverages, tobacco	X
Health	Health costs	
Mobility	Transport costs	X
Social participation	Communications, leisure, services, restaurants and hotels	
Education	Culture and education	
Ecological footprint	Nutrition, mobility	X
Water footprint	Water	X

Source Authors 2020 based on UNDP [68]

functionings of the ecological and water footprint summarizing the worldwide reach of the households. The capability budget of German households contains direct and indirect functionings, which are used to describe the indirect and direct effects of the social life in Germany beyond the classic economic welfare analysis. In the following, the distribution of the capability budget is analysed against the background of Sen's justice theory.

4.2 German Capability Index (GCI) of the Average Household

Like the HDI, the Capability Index is a composite indicator which covers the following dimensions of human welfare of households: welfare basis, decent life, nutrition, mobility, social participation, education and health. The aim is to make the capabilities of the various German households comparable and establish the CI as an index of revealed capabilities.

Based on the data and functionings of Sect. 3, the Capability Index is defined as follows:

First we normalize each direct expenditure and the indirect footprints of the households between 0 and 1 on the basis of the following equation [3, 72]. We set the households in relation to each other:

$$\text{Dimension index} = \frac{\text{actual value} - \text{minimum value}}{\text{maximum value} - \text{minimum value}}$$

We define the maximum and minimum expenditure levels over all income groups of all households for every specific expenditure group:

$$X_{\text{Min}} = \text{Min}\{x(i, j, k)|i \text{ in } I, \ j \text{ in } J, \ k \text{ in } K\}$$
$$X_{\text{Max}} = \text{Max}\{x(i, j, k)|i \text{ in } I, \ j \text{ in } J, \ k \text{ in } K\},$$

i (functionings) $= i = 1...I = 9$, j (income class) $= j = 1...J = 9$), k (household type) $= k = 1...K = 6$.

In the second step, we calculate A^1:

$$A^1_{i,j,k} = \frac{\ln(x_{i,j,k}) - \ln(X_{\text{Min}})}{\ln(X_{\text{Max}}) - \ln(X_{\text{Min}})}.$$

We use the logarithm of the expenditures, because achieving a respectable welfare level of human development does not require unlimited income and expenditures [71], as is shown by the Easterlin paradox [15, 65]. The paradox summarizes the fact that an increase in income is positively correlated with an increase of individual benefits only up to a specific level of income; thus, above a certain threshold an improvement of the income situation is no longer connected to a similar increase of the benefit level.

Hence, we receive the Capability Index ($\text{GCI}_{i,k}$) of the specific income class ($j = 1...9$) of the analysed household type of the social group ($k = 1...6$) adjusted by the specific household size (H_k) of the social group.

$$\text{CI}_{i,j,k} = \frac{A^1_{i,j,k}}{H_k}$$

And the aggregate GCI—the capability budget of the households—is defined as:

$$\text{GCI}^k = \sum_{j=1}^{9} \sum_{i=1}^{9} \text{CI}_{i,j,k}$$

The higher the GCI is, the more capabilities the households can achieve. The GCI is defined between 0 and 1. The societal goal is a high GCI. In the case of the indirect capabilities of the ecological and water footprints the social goal is, as before, to minimize the footprint of the households. To convey this goal also to the ecological and water footprints, the equation has to be adjusted.

For the indirect capability index covering the ecological and water footprint of the FEW-nexus A^2 is defined without logarithm because the damage increases continuously without decreasing marginal ecological damage:

$$A^2_{i,j,k} = \frac{(x_{i,j,k}) - (X_{\text{Min}})}{(X_{\text{Max}}) - (X_{\text{Min}})}$$

The GCI for the FEW-nexus is:

$$\text{GCI}_{i,j,k}^{\text{FEW}} = 1 - \left(\frac{A_{i,j,k}^2}{H_k} \right)$$

The lower the water footprint and the ecological footprint are, the better is the global ecological and water situation caused by the consumption patterns of the German households and the higher is the GCI^{FEW} index.

Based on these basic equations, we calculate the German CI for the different household groups and the various income groups. In the following, we present the first results of our analysis.

4.3 Operationalization of Sen's Justice Approach—The Justice Spread

The functionings of the households are summarized in their capability budgets. This budget thus represents the means of a household to enable a certain life for its members, or—in other terms—the "worldwide reach" of the household. The direct functionings, for example, enable the employed household members to achieve a certain income. This income then enables further functionings, e.g. the next generation through better education, or the current generation through social participation. Hence, the functionings enable both intra- and intergenerational social participation.

However, in the following also the indirect effects of the functionings of the German households are measured through the ecological and water footprint. The actual, practical realization of the functionings of the German households affects the functionings of, for example, those people who are put under pressure through increasing ecological and water footprints. The worldwide reach of the functionings has external spatial effects.

These effects have impacts that can be explained on the basis of Sen's justice approach. For Sen, justice primarily relates to how humans actually live [30] and "how institutions organize the living conditions [21]", or as Rawls set it: "Natural distribution is neither just nor unjust, nor is it unjust that persons are born into society at some particular position. These are simply natural facts. What is just and unjust is the way that institutions deal with these facts (Rawls 1971)". Hence, it is important how the institutions deal with injustice.

Accordingly, the proposed justice spread can be interpreted as an operationalization of Sen's justice theory [21, 59] measuring the effects of institutional behaviour on the distribution of the capability budget between the social groups of German society as well as their impact on other people and societies. In this sense, the justice spread also measures the differences in the worldwide reach between the German households.

The justice spread is defined as the difference of the capability budget between the consumption patterns of the highest and the lowest income group. In the first step, we analyse the capability budget of all households shown in Table 8.

Table 8 Capability budget of all German households—2018

Capability budget of the German households and the FEW-nexus—2018
Monthly expenditures for the households functionings and its external effects
Monthly net income

Social functionings\social groups	Households, total	Below 900	900–1300	1300–1500	1500–2000	2000–2600	2600–3600	3600–5000	5000–18,000	Max	Min	Justice spread
	All households											
Welfare foundation	4047	662	1119	1427	1786	2331	3198	4643	8972	8972	662	14
Nutrition	360	170	193	220	249	295	354	434	540	540	170	3
Decent life[a]	1167	450	569	661	779	963	1260	1521	2034	2034	450	5
Health	50	7	16	18	28	41	58	82	117	117	7	17
Mobility	379	62	101	119	172	238	336	476	762	762	62	12
Social participation	654	170	237	301	367	477	608	789	1184	1184	170	7
Education	28	8	6	7	11	14	21	33	63	63	6	11
External FEW-nexus related effects												
Ecological footprint in m^2	8000	2666	3408	4000	4808	5964	7547	9621	13,778	13,778	2666	5.2
Water footprint in cbm	234	39	55	76	99	140	210	338	585	585	39	15.0

Source Own calculation 2020 based on German Federal Statistical Office (2020), [a]Includes hot, cold and waste water costs; IEK-STE 2020

Table 8 shows the expenditures of German households for the seven functionings chosen for this analysis. The average German household has 4047 € as its welfare foundation and, given this economic basis, spends on average 360 € for nutrition, 1167 € are necessary to finance a decent life, 50 € are spent for health-related issues of the household members, 379 € for mobility, 654 € for the social participation of the members of the household, and 28 € for education. The direct FEW-nexus expenditures of German households are summarized in the functionings nutrition, decent life and mobility. The external FEW-nexus related effects of the consumption patterns of German households are summarized in the ecological and water footprints.

The justice spread between the maximum and the minimum expenditures of the analysed households shows that the welfare foundation of the average German households differs significantly. The households of the highest income group hold a 14 times higher welfare foundation (earnings, savings) than the average household in the lowest income group. The high spread is caused mainly by the fact that the savings of the lowest income group are negative.

Further analysis shows that the spread differs significantly between the functionings. The justice spreads for nutrition and decent life are relatively low (at 3 and 5, respectively). It increases to 7 and 11 for social participation and education and reaches its highest values for health (17) and mobility (12). Hence, the direct FEW-nexus spread ranges from 3 for nutrition to 12 for mobility. Hence, the basic needs are more evenly distributed than, for example, mobility. The spread of the external effects of the households' consumption patterns is inconsistent. The spread of the ecological footprint is relatively low at 5.2, but is significantly higher for the water footprint (15)—the highest income group uses 15 times more direct and indirect water than the lowest income group. Accordingly, the worldwide reach is in this case 15 times higher than that of the low income households in this social group.

The analysis of single woman German households reveals a slightly different picture as Table 9 shows.

Table 9 shows that the justice spread of the welfare foundation is 16, i.e. the justice spread of the single woman households is higher than that of the average household. The distribution for nutrition (2) and decent life (3), in contrast, is lower. The justice spread increases significantly to 15 for health expenditures of single women. Single women in the highest income group can spend on average 15 times more on their health care than the women in the lowest income group. The mobility spread (8) of single woman households are lower than that of the average household. The same holds true for social participation and education (both 6). The table further shows that the justice spread of the external FEW-nexus related effects (4.3) is nearly as low as the spread of nutrition and decent life. So overall it can be concluded that the worldwide reach of the household functionings in this social group is rather heterogeneous.

The average welfare foundation spread of single men is higher (16) than that of the average household and equal to the single woman households (Table 9), whereas the nutrition spread (1.5) of this group is the lowest of all functionings of all analysed households. The justice spread for the decent life is in the range of that of the single women. The spread of the health expenditures, in contrast, is higher (23) than that of

Table 9 Capability budget of single women—2018

Capability budget of the German households and FEW-nexus—2018
Monthly expenditures for the households functionings and its external effects
Monthly net income

Social functionings\social groups	Households, total	Below 900	900–1300	1300–1500	1500–2000	2000–2600	2600–3500	3600–5000	5000–18,000	Max	Min	Justice spread
	Single woman											
Welfare foundation	2207	711	1117	1432	1771	2401	3314	4903	11,477	11,477	711	16
Nutrition	209	159	179	196	210	227	236	259	275	275	159	2
Decent life[a]	824	465	585	664	784	924	1125	1338	1619	1619	465	3
Health	72	26	33	42	55	71	109	181	393	393	26	15
Mobility	173	53	82	112	157	205	284	393	410	410	53	8
Social participation	407	173	228	322	374	472	605	740	1065	1065	173	6
Education	8	7	5	5	7	8	10	12	29	29	5	6
External FEW-nexus related effects												
Ecological footprint in m²	5006	2615	3290	3970	4689	5642	7012	8642	11,216	11,216	2615	4.3
Water footprint in cbm	73.26	38	48	58	69	83	103	126	164	164	38	4.3

Source Own calculation 2020 based on German Federal Statistical Office (2020). [a]Includes hot, cold and waste water costs; IEK-STE 2020

single women, as well as in the case of mobility (11). For social participation (6), in contrast, the spread is the same as that of the single women group. Striking is in the case of the single man households that the spread of the education expenditures with 1.3 is the lowest across all households, as Table 10 shows. The average single man of the highest income group spends only 4 € more for education than the average single man in the lowest income group. The FEW-nexus justice spread of the single man is a little bit lower for the nutrition functioning in comparison with the single woman. The decent life spread is the same, whereas the mobility spread is higher for single men. The external effects of the consumption patterns of the single man households are more independent from the earned income. The effects are just 4 times higher in the highest income group than in the lowest income group. The worldwide reach differs only marginally between the households with rising income.

In the following, we will analyse the capability budget of single parent households.

Table 11 reveals some differences in the capability budget of the **single parents in comparison with the previous households**. The spread of the health (16) and mobility (6) functionings of the single parent households are lower than that of the single households. This also applies to social participation and to some extent for education. In the case of education for the two lowest income groups, no valid data are available. The spread for the nutrition and decent life is as low as for the single households, but the mobility spread is significantly lower. The mobility needs of this group are similar, independent of the earned income. This also applies to the external FEW-nexus related effects. The spread for the ecological footprint (3.4) and for the water footprint (4.5), i.e. the worldwide reach, shows no significant differences between the households according to their earned income.

Table 12 reveals the capability budget of the **couple households without children**, where the analysis reveals a more evenly distributed welfare foundation. The justice spreads for all functionings are below 10, constituting the lowest spread of all analysed households. The highest spreads are identified for both the welfare foundation and health expenditures (both 9). Nutrition (2) and education (2) are the most evenly distributed functionings for the couples.

Similarly, also the FEW-nexus related expenditures of this household group are below 10. The highest FEW-nexus spread is identified for the mobility functionings (7). The external FEW-nexus related effects show a spread of 3 for the ecological footprint and of 4.2 for the water footprint. Hence, the external effects are not so much influenced by the level of the household income.

In a next step, Table 13 shows the capability budget of **couples with children**. On average, for these households the expenditures for the functionings are also more evenly distributed. Only the health expenditures differ significantly between the highest and the lowest income group. The justice spread is 12.

For all other functionings, the justice spreads are below 10 and, again, nutrition is the most evenly distributed functioning (2), followed by the ecological footprint spread (2.7) and the water footprint spread (3.8). Also, the other FEW-nexus related functionings show a relatively low spread between the highest and lowest income group. Hence, these social groups have more similar household needs, which are more independent from the earned income as in other social groups.

Table 10 Capability budget of single men—2018

Capability budget of the German households and FEW-nexus—2018
Monthly expenditures for the households functionings and its external effects
Monthly net income

Social functionings\social groups	Households, total	Below 900	900–1300	1300–1500	1500–2000	2000–2600	2600–3600	3600–5000	5000–18,000	Max	Min	Justice spread
	Single man											
Welfare foundation	2672	655	1059	1449	1892	2507	3483	5082	10,153	10,153	655	16
Nutrition	217	175	193	204	222	229	234	242	261	261	175	1.5
Decent life[a]	769	447	545	627	683	814	964	1144	1327	1327	447	3
Health	71	15	24	30	43	53	83	160	352	352	15	23
Mobility	245	48	118	141	195	277	380	466	535	535	48	11
Social participation	439	160	262	319	398	495	504	771	944	944	160	6
Education	8	7	7	6	7	9	9	9	11	9	7	1.3
External FEW-nexus related effects												
Ecological footprint in m²	5115	2494	3334	3817	4473	5426	6618	8092	10,077	10,077	2494	4.0
Water footprint in cbm	75	37	49	56	65	79	97	118	147	147	37	4.0

Source Own calculation 2020 based on German Federal Statistical Office (2020). [a]Includes hot, cold and waste water costs; IEK-STE 2020

Table 11 Capability budget of single parents—2018

Capability budget of the German households and FEW-nexus—2018

Monthly expenditures for the households functionings and its external effects

Monthly net income

Social functionings\social groups	Households, total	Below 900	900–1300	1300–1500	1500–2000	2000–2600	2600–3600	3600–5000	5000–18,000	Max	Min	Justice spread
Single parents												
Welfare foundation	2738	Null	1195	1419	1795	2313	3836	4678	8304	8304	1195	7
Nutrition	333	Null	232	257	305	339	352	422	321	422	232	2
Decent life[a]	928	Null	566	629.5	748	977.5	1187	1478	1981	1981	566	4
Health	60	Null	18	22	28	41	57	119	283	283	18	16
Mobility	226	Null	172	109.5	211	305.5	314	522	690	690	109.5	6
Social participation	523	Null	233	251	347	510	660	837	896	896	233	4
Education	41	Null	Null	Null	28	38	47	91	91	91	28	3
External FEW-nexus related effects												
Ecological footprint in m²	6533	Null	3811	3923	5006	6213	7364	10,083	13,050	13,050	3811	3.4
Water footprint in cbm	229	Null	123	132	176	218	259	369	554	554	123	4.5

Source Own calculation 2020 based on German Federal Statistical Office (2020). Null = not enough data available. [a]Includes hot, cold and waste water costs; IEK-STE 2020

Table 12 Capability budget of couple households without children—2018

Capability budget of the German households—2018
Monthly expenditures for the households functionings and its external effects
Monthly net income

Social functionings\social groups	Households, total	Below 900	900–1300	1300–1500	1500–2000	2000–2600	2600–3600	3600–5000	5000–18,000	Max	Min	Justice spread
	Couples without children											
Welfare foundation	4962	Null	1015	1300	1794	2358	3140	4594	8790	8790	1015	9
Nutrition	415	Null	299	315	339	363	400	427	474	474	299	2
Decent life[a]	1341	Null	Null	755	817	977	1290	1522	1963	1963	755	3
Health	173	Null	46	35	53	84	108	163	329	329	35	9
Mobility	442	Null	198	115	178	268	478	600	820	820	115	7
Social participation	807	Null	283	278	366	509	638	839	1246	1246	278	4
Education	11	Null	8	10	13	9	9	11	15	15	8	2
External FEW-nexus related effects												
Ecological footprint in m^2	9435	Null	4683	4453	5269	6432	7947	9796	13,491	13,491	4453	3.0
Water footprint in cbm	359	Null	137	130	162	217	279	401	553	553	130	4.2

Source Own calculation 2020 based on German Federal Statistical Office (2020). Null = not enough data available. [a]Includes hot, cold and waste water costs; IEK-STE 2020

Table 13 Capability budget of couples with children—2018

Capability budget of the German households—2018
Monthly expenditures for the households functionings and its external effects
Monthly net income

Social functionings\social groups	Households, total	Below 900	900–1300	1300–1500	1500–2000	2000–2600	2600–3600	3600–5000	5000–18,000	Max	Min	Justice spread
	Couples with children											
Welfare foundation	6639	Null	1012	1246	1808	2388	3220	4684	8752	8752	1012	9
Nutrition	536	Null	325	357	392	449	474	518	585	585	325	2
Decent life[a]	1588	Null	285	296	821	927	1272	1572	2093	2093	285	7
Health	117	Null	18	15	30	42	51	79	174	174	15	12
Mobility	588	Null	265	195	191	335	513	587	876	876	191	5
Social participation	900	Null	306	285	319	447	572	768	1167	1167	285	4
Education	104	Null	Null	16	48	50	71	87	133	133	16	8
External FEW-nexus related effects												
Ecological footprint in m²	11,337	Null	Null	Null	5263	6731	8030	9962	14,080	14,080	5263	2.7
Water footprint in cbm	18,916	Null	Null	Null	6471	10,343	12,692	16,620	24,728	24,728	6471	3.8

Source Own calculation 2020 based on German Federal Statistical Office (2020). Null = not enough data available. [a]Includes hot, cold and waste water costs; IEK-STE 2020

The analysis so far has revealed heterogeneity in the capability budget across the household groups and their respective justice spreads. Hence, the different effects will now be summarized in the capability index for the various German household groups based on the equations of Sect. 4.2.

4.4 Revealed Capabilities

The analysis reflects the heterogeneous picture of the German capability index (GCI) for the various households (Table 14). The GCI measures the distribution of the capability budgets between the social groups and their functionings.

Our results reveal a great spread between the social groups over all functionings. The highest GCI is calculated for the group of single women and single men (0.469, 0.458), followed by the group of couples (0.376), and at a much lower level the single parents (0.274) as well as couples with children (0.164). These numbers result in an aggregated GCI for the average household of 0.248. The capability budget of the single parents and of the couples with children is smaller than that of the single households. Essentially this implicates that a single woman or a single man can make much better use of their functionings than members of the other household groups.

Table 14 German capability index

German capability index 2018 and the FEW-nexus
Per person of household—adjusted by household size

Functionings	All households	Single woman	Single man	Single parents	Couples without children	Couples with children
Welfare	0.318	0.424	0.491	0.210	0.354	0.214
Nutrition	0.314	0.210	0.239	0.238	0.368	0.247
Decent life	0.348	0.551	0.486	0.237	0.374	0.217
Decent life energy	0.237	0.304	0.271	0.190	0.274	0.155
Health	0.244	0.579	0.575	0.224	0.398	0.185
Mobility	0.342	0.480	0.582	0.238	0.383	0.227
Mobility energy (fuel)	0.282	0.420	0.516	0.217	0.353	0.214
Social participation	0.311	0.404	0.443	0.228	0.351	0.206
Education	0.341	0.425	0.425	0.319	0.245	0.251
Over all functionings	0.248	0.469	0.458	0.274	0.376	0.164
FEW-nexus $(1 - X)$	Indirect German capability index 2018					
Ecological footprint	0.525	0.597	0.585	0.444	0.231	0.125
Water footprint	0.353	0.776	0.770	0.172	0.133	0.023

Source Own calculation 2020; IEK- STE 2020

In the following, we analyse the six major groups of functionings. In the case of the **welfare functionings**, the single man has the highest index value (0.491), followed by the single women (0.424) and couples (0.354). The single parents (0.210) and couples with children (0.214) have the lowest index values. Given this data, we can conclude that the welfare basis of families with children is lower than that of single households and couples.

When look specifically at the **decent life functioning**, it shows that the social group of single women has the highest value (0.551), followed by couples without children (0.374) and single men (0.486). Families with children have significantly lower index values. Households with children have difficulties in achieving the same level of decent life as households without children. Their capabilities are limited by their functionings. When looking at the health functioning, three major groups of households are identified. The CGI of single households is more than twice that of the households with children and of couples without children.

As discussed above, the FEW-nexus can be characterized by the energy, mobility and nutrition functionings.

The analysis of the **functioning nutrition** reveals that couples without children have the highest value (0.368) and the single woman households have the lowest value (0.210), whereas the other household groups have more or less similar values. The **energy functionings** for a decent life show a similar distribution as the decent life results, but the values are on average a little lower. The spread between the highest and the lowest value is smaller; i.e. the energy capability budget is more evenly distributed.

Also for the **mobility functioning**, a very heterogeneous picture of German households is revealed. In the case of mobility, the single man households have the highest GCI value (0.516) followed by single women and couples. The index is significantly lower for families with children, implying that these families on average have less mobility options than the other households.

This has further implications for the **social participation** options **of the households**. Single men, single women and couples have a significantly higher opportunity to use their capabilities in the social participation process. Families, in contrast, have limited social participation options as indicated by their low CGI. The data show that families generally save their capabilities for the **education functioning**. Their values are much higher than those of the single households. They focus their functionings to provide their children better capabilities and a better life in the future. Accordingly, in their capability budget the functioning education has a significantly higher value than the other functionings. Our finding confirms the central assumption of Sen that the capability approach includes explicit value judgements which differ from person to person [14, 57] and can be revealed by the GCI.

The analysis of the indirect German capability index shows that the impact of the single households on the local and global water resources is much lower than of the other households, as reflected in the highest CGI for the water footprint by single women (0.776) and single men (0.770). The single parents and the couples have significantly lower GCIs, so that the water footprint GCI of the average German household is 0.353. This average household value is lower than the CGI value for the ecological footprint (0.525). That is, for the utilization of the general ecological resources the GCI values are lower for the average households than for the usage of the local and global water resources.

The differences between the various households of the analysed social groups become smaller when considering the total environmental impact of the households' consumption patterns. The GCI of single households is very similar between women and men (0.597, 0.585), whereas for the single parents (0.444), couples with (0.125) and without (0.231) children, the index decreases, resulting in an overall average household value of 0.525. In summary, these numbers show that the average impact of the households on the environment is lower than on the global water system.

5 Conclusion—The Worldwide Reach of German Households

We showed that functionings are constitutive of a person's well-being [58]. Furthermore, we analysed the functionings of German household groups. The presented analysis shows that people use their functionings differently to realize their capabilities and that they have distinct capability budgets. As expected, the budget of households with children is more limited than that of single households and couples. Family households focus their functionings more strongly. They have a clear priority for education, i.e. on the future well-being of their children. We assume that families restrict their capabilities in some areas, as for example mobility and health, and focus more on education. The high values for health in the groups of single man and woman households is also caused by the fact that a large percentage of these households are retirees with necessarily higher health expenditures.

The analysis has also shown that the worldwide reach of the German households can be measured and that the impacts on the local and global water and ecological resources increase with growing income. The higher the income of the households the higher is the utilization and exploitation of global resources.

Sen's welfare concept expressed in his capability approach can be related to the global quest for sustainable development [41]. The presented analysis has shown that a capability-based sustainable development concept has to consider also the spatially distributed external effects imposed on the current generation around the globe. These effects influence the functionings of the people who provided the water and ecological resources of the German household consumption patterns. The analysis confirms the initially discussed assumption that society (analysed according to its household groups) and its households have to increasingly sustain itself as the ecological and water footprints become bigger and bigger with rising household income.

Hence in the sense of Kant and Rosa, our analysis reconfirms two important normative restrictions:

1. The respect for the capabilities and functionings of other people should be an objective law, the own consumption patterns affect other people's capabilities and functionings.
2. Act at any time so that your world in reach does not become continuously larger [55] at the expense of others.

Furthermore, the developed analytical approach systematically proved the immediate intra- and intergenerational connection between individual well-being and the pursuit of sustainable development.

6 Outlook on Further Research Needs

Further research is needed to analyse the reasons for the continuously growing worldwide reach of the German households with rising income and social status. It should be analysed, to what extent the ecological and water footprints of the German households and their respective impacts on the FEW-Nexus sectors can be reduced through alternative consumption patterns or through institutional and governmental measures. Hence, more research is needed on the institutional implementation of the revised Hartmut Rosa imperative: Reduce the worldwide reach of the German households and their consumption decisions without reducing the utility level of the German households. This research should include the whole value added chain of the production and consumption processes: From cradle to grave all relevant economic processes should be identified and captured.

Another research step is the dynamization of the German capability index model, to generate an image of the future worldwide reach of German households against the background of different institutional, political and economic measures set by the decision-makers. This would enable a simulation of different future pathways initiated by the German institutions.

More in-depth research is needed on the management opportunities of the FEW-Nexus in accordance with sustainable development, thus without restraining the capabilities of current and future generations.

References

1. Altman M (2012) Sen's "capabilities" and economic welfare. Elsevier Science, Burlington
2. Anand P, van Hees M (2006) Capabilities and achievements: an empirical study. J Soc Econ 35:268–284
3. Anand S, Sen A (1994) Human development index: methodology and measurement. Occacional papers. UNDP, New York
4. Anand S, Sen A (2000a) Human development and economic sustainability. World Dev 28:2029–2049
5. Anand S, Sen A (2000b) The income component of the human development index. J Hum Dev 1:83–106

6. Australian Government Department of Health (2020) Water—its imprtance and source. Australian Government. https://www1.health.gov.au/internet/publications/publishing.nsf/Content/ohp-enhealth-manual-atsi-cnt-l~ohp-enhealth-manual-atsi-cnt-l-ch6~ohp-enhealth-manual-atsi-cnt-l-ch6.1. Accessed 5 Oct 2020
7. Ayres RU, van den Bergh JCJM, Gowdy JM (1998) Viewpoint: weak versus strong sustainability. Tinbergen discussion paper 1998
8. Ayres RU, van den Bergh JCJM, Gowdy JM (2001) Strong versus weak sustainability: economics, natural sciences, and "consilience". Environ Ethics 23:155–168
9. Bartiaux F, Schmidt L, Horta A, Correia A (2016) Social diffusion of energy-related practices and representations: patterns and policies in Portugal and Belgium. Energy Policy 88:413–421. https://doi.org/10.1016/j.enpol.2015.10.046
10. Bartiaux F, Vandeschrick C, Moezzi M, Frogneux N (2018) Energy justice, unequal access to affordable warmth, and capability deprivation: a quantitative analysis for Belgium. Appl Energy 225:1219–1233. https://doi.org/10.1016/j.apenergy.2018.04.113
11. Biggs EM et al (2015) Sustainable development and the water–energy–food nexus: a perspective on livelihoods. Environ Sci Policy 54:389–397. https://doi.org/10.1016/j.envsci.2015.08.002
12. Burns S et al (2006) The ecological wealth of nations. Global Footprint Network, Oakland
13. Clark H (2017) Human development means realizing the full potential of every life. UNDP. https://www.undp.org/content/undp/en/home/blog/2017/3/21/Human-development-means-realizing-the-full-potential-of-every-life.html
14. Deneulin S, Shahani L (eds) (2009) An introduction to the human development and capability approach. Earthscan, London
15. Easterlin RA (1974) Does economic growth improve the human lot? Some empirical evidence. In: David PA, Reder MW (eds) Nations and households in economic growth. Academic Press, pp 89–125. https://doi.org/10.1016/B978-0-12-205050-3.50008-7
16. European Commission (2003) Household budget surveys in the EU methodology and recommendations for harmonisation—2003. EU, Luxembourg
17. FAO (2014) Water—the most basic resource but also the most essential. FAO. https://www.fao.org/zhc/detail-events/en/c/231215/. Accessed 2 Oct 2020
18. Federal Statistical Office (Statistisches Bundesamt) (2013) Economic accounts, income and consumption samples, task, method and implementation (Wirtschaftsrechnungen Einkommens- und Verbrauchsstichprobe Aufgabe, Methode und Durchführung). Fachserie Wirtschaftsrechnungen Fachserie 15
19. Federal Statistical Office Germany (2020a) Wirtschaftsrechnungen. Einkommens- und Verbrauchsstichprobe Einnahmen und Ausgaben privater Haushalte 2018
20. Federal Statistical Office Germany (2020b) Wirtschaftsrechnungen. Einkommens- und Verbrauchsstichprobe Konsumausgaben privater Haushalte 2018
21. Fischer W, Hake J-F (2017) The food, energy, water nexus: global justice as a driving force of integrated and coherent resource governance? In: SDEWES conference 2017, Dubrovnik
22. Fukuda-Parr S (2003) The human development paradigm: operationalizing Sen's ideas on capabilities. Fem Econ 9:301–317
23. German Environmental Agency (Umweltbundesamt (UBA)) (2018) Wasserfußabdruck. UBA. https://www.umweltbundesamt.de/themen/wasser/wasser-bewirtschaften/wasserfussabdruck. Accessed 2 Oct 2020
24. German Federal Statistical Office (2005) The income and consumption survey (EVS)—tasks, methods and implementation (in German:Einkommens- und Verbrauchsstichprobe—Aufgabe, Methode und Durchführung der EVS). Special serie household budget surveys (in German: Fachserie Wirtschaftsrechnungen) 15
25. Global Footprint Network (2019) National footprint and biocapacity accounts, 2019 edn. Global Footprint Network
26. Global Footprint Network (2020) Glossary. Global Footprint Network. https://www.footprintnetwork.org/resources/glossary/. Accessed 18 Oct 2020
27. Gutteres A (2019) Davos speech 2019, January 24, 2019th edn. World Economic Forum, Davos

28. Haq M (1992) Human development in a changing world. UNDP, New York
29. Haq M (1995) Reflections on human development. Oxford University Press, Oxford
30. Heidenreich F (2011) Theorien der Gerechtigkeit: Eine Einführung. Verlag Barbara Budrich, Opladen, p 251
31. Hillerbrand R (2018) Why affordable clean energy is not enough. A capability perspective on the sustainable development goals. Sustainability 10(7):2485. https://doi.org/10.3390/su1007 2485
32. IPCC (2018) Global warming of 1.5°C—summary for policymakers. IPCC, Incheon
33. IPCC (2019a) IPCC special report on climate change, desertification, land degradation, sustainable land management, food security, and greenhouse gas fluxes in terrestrial ecosystems—summary for policymakers. IPCC, Geneva
34. IPCC (2019b) The ocean and cryosphere in a changing climate—summary for policymakers. In: Second joint session of working groups I and II of the IPCC and accepted by the 51st session of the IPCC, principality of Monaco, Monaco, 24 Sept 2019
35. Kant I (ed) (1785) Grounding for the metaphysics of morals. Peter Millican and Amyas Merivale (both of Hertford College, Oxford), Oxford
36. Kitzes J, Buchan S, Galli A, Ewing B, Shengkui C, Gaodi X, Shuyan C (2008) Report on ecological footprint in China. China Council for International Cooperation on Environment and Development (CCICED), WWF China, Peking
37. Krugman P, Wells R (2018) Economics. Worth, New York City
38. Kuklys W (2005) Amartya Sen's capability approach. Springer, Berlin
39. Lessmann O (2009) Conditions of life, functionings and capability: similarities, differences and complementary features. J Hum Dev Capabil 10:279–298. https://doi.org/10.1080/194528 20902941271
40. Leuprecht P (2005) Contraindre le fort pour affranchir le faible. Rev Relat
41. Leßmann O, Masson T (2015) Sustainable consumption in capability perspective: operationalization and empirical illustration. J Behav Exp Econ 57:64–72. https://doi.org/10.1016/j.socec. 2015.04.001
42. Leßmann O (2014) Arbeit und das gute Leben—Erfassung von Verwirklichungschancen im Capability-Ansatz. In (Hg.): Friedrich-Ebert-Stiftung (ed) Was macht ein gutes Leben aus? Der Capability Approach im Fortschrittsforum. Friedrich-Ebert-Stiftung, Bonn, pp S. 47–56
43. Lin D et al (2019) Working guidebook to the national footprint and biocapacity accounts. Global Footprint Network, Oakland
44. Martínez-Guido SI, González-Campos JB, Ponce-Ortega JM (2019) Strategic planning to improve the human development index in disenfranchised communities through satisfying food, water and energy needs. Food Bioprod Process 117:14–29. https://doi.org/10.1016/j.fbp. 2019.06.007
45. Merkel W (2007a) Soziale Gerechtigkeit: Theorie und Wirklichkeit
46. Merkel W (2007b) Soziale Gerechtigkeit: Theorie und Wirklichkeit. https://fes-online-akademiede/startseite/
47. Neumayer E (1999) Weak versus strong sustainability. Edward Elgar Publishing, Cheltenham
48. Neumayer E (2003) Weak versus strong sustainability. Exploring the limits of two opposing paradigms, 2nd edn. Edward Elgar Publishing, Cheltenham
49. Nussbaum M (2007) Human rights and human capabilities. Har Hum Rights J 20:20
50. Nussbaum M, Sen A (eds) (1993) Quality of life. Oxford University Press, Oxford
51. Rawls J (1971) A theory of justice. Cambridge, MA: Harvard University Press
52. Robeyns I (2005) The capability approach: a theoretical survey. J Hum Dev 6:93–117. https://doi.org/10.1080/146498805200034266
53. Robeyns I (2011) The capability approach. Stanford
54. Rosa H (2019) Resonanz: Eine Soziologie der Weltbeziehung. Suhrkamp, Berlin
55. Rosa H (2017) Auf eine andere Art mit der Welt in Beziehung treten. UMWELTPERSPEKTIVEN Der UFZ—Newsletter | Dezember 2017, Leipzig
56. Rosa H (2018) Hartmut Rosa über Resonanz. Resonanz Wien. https://www.resonanz.wien/blog/hartmut-rosa-ueber-resonanz/. Accessed 12 Oct 2020

57. Sen A (1999) Development as freedom. Anchor, New York
58. Sen A (2007) Inequality reexamined. Harvard University Press, Cambridge
59. Sen A (2009) The idea of justice. Penguin Books, London
60. Sen A (2013) The ends and means of sustainability. J Hum Dev Capabil 14:6–20
61. Sen A (1979) Utilitarianism and welfarism. J Philos 76:463–489
62. Sen A (1985a) Commodities and capabilities. Elsevier, Amsterdam
63. Sen A (1985b) Well-being, agency and freedom: the Dewey lectures 1984. J Philos 82:169–221
64. Sen A (2003) Ökonomie für den Menschen [Economy for the people], 2nd edn. dtv, Munch
65. Shin D (1980) Does rapid economic growth improve the human lot? Some empirical evidence. Soc Indic Res 8:199–221. https://doi.org/10.1007/BF00286477
66. Siebert H (1982) Nature as a life support system. Renewable resources and environmental disruption. Z Nationalökon/J Econ 42:133–142
67. Sovacool BK, Dworkin MH (2015) Energy justice: conceptual insights and practical applications. Appl Energy 142:435–444. https://doi.org/10.1016/j.apenergy.2015.01.002
68. UNDP (2007) Measuring human development. A primer. Guidelines and tools for statistical research, analysis and advocacy. UNDP, New York
69. United Nations (2015) Transforming our world: the agenda for sustainable development. United Nations, New York
70. United Nations (UN Water) (2020) Water, food and energy. United Nations. https://www.unwater.org/water-facts/water-food-and-energy/. Accessed 30 Sept 2020
71. United Nations Development Programme (UNDP) (1990) Human development report 1990. Oxford University Press, Oxford
72. United Nations Development Programme (UNDP) (2007) Human development report 2007/2008. UNDP, New York
73. Venkatapuram S (2011) Health justice. Polity Press, Cambridge
74. WWF (2002) Living planet report 2002—summary. WWF, Gland
75. WWF (2004) Living planet report 2004. Banson, Cambridge
76. WWF (2006a) Ecological footprint and biocapacity. Technical notes: 2006 edition. 1. The ecological footprint. WWF, Cambridge
77. WWF (2006b) Living planet report 2006. Banson, Cambridge
78. WWF (2020) Living planet report 2020. WWF, Gland
79. Wackernagel M, Onisto L, Callejas LA, Lopez FS, Mendez G (1997) Ecological footprints of nations. How much nature do they use? How much nature do they have? Universidad Anahuac de Xalapa, Centro de Estudios para la Sustentabilidad, Xalapa
80. Wackernagel M, Onisto L, Lineares A, Falfan ISL, Garcia JM, Guerrero AIS, Guerrero MGS (2002a) Ecological footprints of nations. How much nature do they use?—How much nature do they have? Ecouncil, Mexico
81. Wackernagel M, Rees W (1996) Our ecological footprint reducing human impact on the earth. New Society Publishers, Gabriola Island
82. Wackernagel M et al (2002b) Tracking the ecological overshoot of the human economy. Proc Natl Acad Sci USA 99:9266–9271
83. Water Footprint Network (2020) What is a water footprint. Water Footprint Network. https://waterfootprint.org/en/water-footprint/what-is-water-footprint/. Accessed 10 Oct 2020
84. Wied-Nebbeling S (2007) Grundlagen der Mikroökonomik (Springer-Lehrbuch). Springer, München

Energy, Water, Food Nexus Decision-Making for Sustainable Food Security

Sarah Namany and Tareq Al-Ansari

Abstract Natural resources and their constituting systems continue to experience stresses attributed to expanding anthropogenic activities. Economic expenditures on most sectors are heavily dependent on water, energy and food resources, the utilisation of which contributes to resource depletion and environmental burdens. The energy–water–food (EWF) nexus is a concept that harnesses the intrinsic relationships existing between resources and representative sub-systems in order to support efficient resources management, develop resilience and enhance environmental consciousness. While the EWF nexus framework can provide preliminary decision-making insights into resource sub-systems, it cannot alone address modern-day multidimensional resource problems and their related economic and environmental challenges. Food insecurity, a global challenge for mankind, is a result of various factors, of which is resource mismanagement. This chapter investigates the possibility of understanding and alleviating food insecurity by means of the EWF nexus approach. It highlights using case studies the possible analytical and modelling decision-making tools that enable the characterisation and analysis of resources systems across strategic, tactical and operational levels of decision-making in volatile and risky environments. Example of studies involve optimisation, machine learning and agent-based modelling, and illustrate the complementary nature of these tools within decision frameworks, and how they can be utilised to alleviate food insecurity and support sustainable and resilient resource management.

Keywords EWF nexus · Decision-making · Food security · Resource management · Optimisation · Machine learning · Agent-based modelling

S. Namany · T. Al-Ansari (✉)
Division of Sustainable Development, College of Science and Engineering, Hamad Bin Khalifa University, Doha, Qatar
e-mail: talansari@hbku.edu.qa

T. Al-Ansari
Division of Engineering Management and Decision Sciences, College of Science and Engineering, Hamad Bin Khalifa University, Doha, Qatar

1   Introduction: Food Insecurity Background Information: Trends and Causes

Over the years, the global population and economy has and continues to experience significant demographic expansion, urbanisation and industrialisation [1]. However, the continued increase in population is accompanied by growing demands on resources and industrial services, which in turn cause tremendous pressures on natural resources and the surrounding environment. In fact, the global population is expected to reach 9.7 billion by 2050 [2], leading to a significant surge in demands for all major industrial sectors. Food demand is predicted to rise by 70%, and global energy demand by more than 50%. The water sector which heavily supports agriculture activities, which is the main consumer of fresh water at a share of 70%, will experience unprecedented withdrawal rates to meet the requirements of other sectors [3]. In order to supply sufficient resources to all sectors and to feed the increasing population, resource sectors should expand to accommodate the rise in production. This would cause significant pressures on the environment due to the emissions associated with power generation, water withdrawal or treatment and food production and primary agriculture activities. In addition, pressures on resource sectors and resource depletion are not the only risks associated with demographic expansion. From a social perspective, the increased demand for food products can lead to disparity in the access for food in many regions of the world especially in those with limited financial resources and constrained access to the natural capital. In order to overcome the uneven food situation, the concept of food security is adopted as an adaptation strategy to ensure equal and fair access to food products. Food security is defined as the state *"when all people, at all times, have physical and economic access to sufficient, safe and nutritious food to meet their dietary needs and food preferences for an active and healthy life"* [4]. It is comprised of four main pillars: availability, accessibility, utilisation and stability. Food availability consists of all activities responsible for delivering food products including production and supply chains (distribution). Food accessibility is related to the economic value of food products along with their social affordability. As for food utilisation, it is concerned with food safety and sanitation in addition to the nutritional content of food products. Food stability is not generally considered a pillar on its own, although in modern society it is a critical component to food security as it maintains the stability of the three core pillars [5]. In order to achieve holistic food security, the representative pillars should be modelled efficiently as a means to ensure the continuous performance and effective operations of all subsystems that constitute the food system. It is also important to highlight that besides the fundamental pillars of food security; it cannot be achieved in isolation to the secure provision of water and energy resources. Therefore, modelling food security requires the consideration of the interactions that exist between the three sectors in order to ascertain an accurate representation for resource requirements, synergies, and trade-offs. As such, the energy, water, and food (EWF) nexus is an approach that can enable the efficient cooperation between EWF resources for optimal resource management, environmental protection and enhanced economic performance.

In this chapter, the importance of the EWF nexus in addressing resource management issues and food security will be highlighted. The focus will be on the decision-making tools and techniques that have demonstrated effectiveness in addressing food security, with a major emphasis on the resilience and the risk alleviation component as a means to maintain stability of the food system. First, simulation models will be presented as they can realistically simulate the food sector and associated resource sectors. Then, risk quantification techniques will be discussed to illustrate the importance of modelling uncertainty within resource systems. Finally, optimisation models will be explained as a technique that utilises the output of the previous methods to enhance the performance of the overall food sector and supporting resources systems. Collectively, these tools can support informed decision-making related to food security and can be tailored to specific cases across various temporal and spatial applications and case studies.

2 Resources Management: Decision-Making Challenges

Food security, with all its associated pillars, is directly dependent on the security of water and energy sectors. The availability of food, including domestic food production along with national and international supply chains, requires continuous resource inflows from water and energy systems for regular operations. In fact, water withdrawal for agricultural activities accounts for 70% of the total freshwater extracted globally, with an expected increase in the coming years to account for rising food demands. This surge is estimated to reach 60% by 2050 to meet the ever-growing population [6]. Similarly, accessibility of food products is highly influenced by the affordability of water and energy resources, and any volatility in the prices of these two commodities can disturb the food market prices. A study conducted by Taghizadeh-Hesary et al. has proven a significant correlation between oil and food prices, such that a change of 64.7% in food prices was recorded when oil prices were fluctuating between 2000 and 2016. Results from this study advise the need to diversify the energy mix in the agricultural sector, by adopting renewables and reducing the total reliance on non-renewable energy sources, to maintain a secure access to energy and reduce the influence of volatility of fossil fuels on food prices [7]. In addition, the stability and resilience of the food system, a crucial element for the achievement of a holistic food security, is significantly linked to the availability and stability of other systems. For instance, food cultivation necessitates a continuous access to arable lands and irrigation water to grow crops [8]. Similarly, the daily operations of livestock farms are dependent on electricity [9], as it is needed for the cooling of stables of lactating cows [10] and poultry; the storage of produced food in the case of milk production [11]; and the drying of crops for animal feed and other cultivated crops [12]. In regions characterised by arid and hot climate conditions, growing food in the fields is extremely challenging due to the extremely high temperatures and inadequate soil quality. Greenhouses are an alternative cultivation technology that have gained a substantial importance over the last few years, attributed to their capability to accommodate optimal growing conditions for crops such that cultivation is feasible throughout the year, even under the most severe climates [13]. The inherent interlinkages between resources between the food system and the energy and water systems are collectively advantageous for the three resources sectors as they represent an opportunity for fruitful cooperation. However, such interdependencies can also be the impetus for various tradeoffs and competing objectives, which can affect the performance of the food system and hence obstruct targets, for instance food security. Naturally, multi-sectoral operations which are a consequence of interlinkages between resources imply the participation and cooperation between multiple stakeholders, which in most cases demonstrate divergent objectives and targets. Multi-stakeholders accompanied with their multiple objectives induce intersectoral and intra-sectoral competitions that perturb decision-making within the food system. An example of intersectoral competition is manifested within competition for water resources between the food system which requires irrigation water to maximise its crops production, and the energy sector that requires water for power generation in power plants [14]. As for the intra-sectoral competition within the food sector, it is visible in the end use of the food products. Food is generally produced to feed people; however, in some instances, agricultural products can be redirected towards the production of electricity, in the form of biomass feedstock used to generate power by means of gasification. This multiple usage of agricultural products can be problematic for the food sector if objectives are not balanced amongst stakeholders [15].

In addition to the decision-making challenges that stem from the contribution of the energy and water systems in the functioning of the food sector, the economy within which the three sectors are operating also imposes a range of rules and policies that can restrict some processes and inhibit the achievement of desired targets within resource sub-systems. Carbon pricing, which is a tax imposed by several governments to minimise greenhouse gas (GHG) emissions associated with industrial activities as means to address climate change, is a policy that represents a burden for several sectors, including those directly and indirectly related to energy, water and food systems [16]. While the ultimate aim of these economic sectors is to maximise their profit and reduce their expenses, the carbon tax is an added financial burden to operations. To avoid paying this tax, some industries invest in alternative environmentally friendly technologies that generate less GHGs, such as solar energy or biomass-based energy production for the energy sector [17], which usually require large capital investments. In the case of carbon pricing, the extra costs associated with emissions generated or technologies deployed imply a reform in the decision-making scheme of these sectors in order to balance the sector's economic objective and the government environmental policy. Another example illustrating the impact of the involvement of multiple stakeholders on the performance of resources is the land constraint imposed by several governments regarding land usage. Such regulations can be a cause for tension between sectors, as it can hinder the productivity of a sector at the expense of another. With the need to reduce the environmental burden associated with conventional fossil fuel-based power generation, bioenergy is used as an alternative energy source owing to its relatively low emissions and implication for carbon neutral power generation. Bioenergy requires crops as the main source of energy, which threatens the productivity of the food sector by delimiting its usage of land for food production [18].

The multi-dimensional aspect of the food system and the multi-stakeholder nature governing its operations create competitions and conflicts due to overlapping interests and targets. This competition, if not turned into cooperative alliances, can reduce the efficiency of the food sector along with its associated subsystems. Furthermore, the food sector with its major dependence on the energy and water resources is a pocket of instability and subject to a range of risks and uncertainties that exacerbate the level of food insecurity. Risks associated with the interconnections imposed by the contribution of different parties are considered internal risks inflicted by system-level disturbances. On the other hand, there exist other external factors that influence the performance of the food sector and its desire to develop holistic food security. The rapid population growth rate, which is expected to raise the global world population to 9.8 billion by 2050, is causing a significant surge in the demand for food products [2]. This increase implies the need to expand and intensify agricultural activities, thus inflicting pressures on the food sector and supporting energy and food technology driven sub-systems. To equally and fairly support and feed the growing population and to strive towards global food security, food production must increase by approximately 70% [19]. Such expansion can significantly pressure the natural resource base, including those related to land, water and energy resources, which in many areas of the world are stressed beyond sustainable capacities [20]. Together with this agricultural shift, environmental emissions are expected to rise due to energy intensive technologies required to maintain continuous food supply [21]. The culmination of these effects gives rise to several sustainability concerns that can jeopardise the efficiency of the food sector. It is also worth noting that impact of population growth on food security differs from one country to another. Due to global disparities related to the distribution of cultivated lands used for cropping purposes, which are generally centered in high-income countries with a capacity of cultivation approximating 0.37 ha per capita, versus 0.23 ha per capita and 0.17 ha per capita for middle and low-income countries [22], agriculture expansion in low and mid-income countries is hardly to be executed aggravating the food insecurity in those countries.

As part of the external challenges governing food security and resources systems, climate change and its incumbent natural disasters is a cause for concern. Climatic conditions such as the temperature, rainfall, and soil quality are critical factors amongst many that govern the productivity of the agricultural sector. Crops and live-stock require specific climatic conditions to survive and grow efficiently, for which any adverse changes in the climate can lead to agricultural losses, and consequently result in shortages in the food available for consumption [23]. Today, the issues associated with climatic conditions can be resolved by implementing advanced technologies such as greenhouses, vertical farming, hydroponics that can mimic optimum natural and cultivation conditions for crops and other products [24, 25]. As for the raising of livestock, establishing cooled or shaded shelters can alleviate the risk of temperature variations and extreme climates [26]. While predictable climate risks can be mitigated by the aforementioned adaptation techniques, natural disasters which are difficult to envisage can expose vulnerabilities and pose a threat to the stability of food systems and resource sub-systems. For instance, in turkey, the extreme droughts that occurred in 2007 resulted in a significant decrease in agricultural production resulting in a 15.3% reduction in crop yields [27]. According to the United States Department of Agriculture (USDA), Hurricane Harvey that impacted the states of Texas and Louisiana in 2017 caused drastic damage to the railway infrastructure leading to supply chain disruptions that affected regional grain exports [28]. Pandemics, which are not necessarily only a result of natural disasters, are a source for unpredictable risks and may result in similar catastrophic impacts on the food sector as with the case of conventional environmentally driven natural disasters. In fact, the recent Covid-19 pandemic has engendered an unforeseen global recession impacting all vital economic sectors and has been cause for much suffering. While some regions of the world suffer from famine and undernourishment due to poverty and other socio-political issues, in many cases the spread of the virus has further undermined food systems, supply chains and ultimately food security. City-wide lockdowns, coupled with movement and transportation restrictions between countries and cities, have engendered disruptions in food supply chains threatening the availability, stability, and accessibility of food products. Furthermore, economic and social outcomes as a result of lockdowns and economy shutdowns have an impact on the purchasing power, and income of individuals due to health complications and safety protocols that have reduced manpower along the supply chain. Expectedly, there are impacts on the utilisation pillar of the food security as the access to nutritious food would be more difficult [29, 30].

Within the myriad of internal and external challenges that can affect processes within EWF resource sub-systems, there is an imminent need to develop strategies and flexible decision-making schemes that can reduce their impact and reduce uncertainties. Positively harnessing the interlinkages existing between the food, water, and energy systems in the context of the EWF nexus system approach can support holistic understanding of the weaknesses and strengths within each system and promote collective system performance enhancements. Section 3 provides an overview of the EWF nexus approach as an analytical tool and its ability to mitigate the previously highlighted challenges focusing on the holistic nexus system. Section 4 details specific modelling challenges that arise when solving resources management problems, along with insight into the methods that can overcome these challenges, with a major focus on food security cases.

3 EWF Nexus Approach: A Tool to Harness Resources Interactions in Achieving Sustainable Resources Management

The EWF nexus is a systems approach that emphasises on the intricate interactions between energy, water and food resource sub-systems [31], first introduced in the Bonn conference in 2011 as means to address the unsustainable resource management across EWF sectors [32]. The Bonn conference defined the pillars constituting the security of each resource system as part of defunding relationships between systems, and to encourage sustainable and rational quantification of interactions amongst them and their associated subsystems [33]. In this regard, energy security was defined as the consistent access to cheap and uninterrupted energy for the different daily-life activities [34]. Water security is described as the safe and sufficient access to water resources that ensure a decent livelihood [35]. As for food security, it is represented by four pillars that ensure a consistent and stable availability, affordability and utilisation of food resources [4]. Considering the security of security of resources, ultimately the EWF nexus framework establishes an equilibrium between diverse system objectives by balancing synergics and tradeoffs, while promoting sustainability and maintaining ecosystem integrity [36].

While the EWF nexus approach is generally adopted as a tool to ensure efficient resources management, it can serve different purposes depending on the context in which it is applied in. According to Albrecht et al. the EWF nexus can be considered as an analytical tool to quantitatively or qualitatively describe interactions between sectors, using mathematical models or social analytical frameworks [37]. This can be applied to different levels of the decision making ranging from the operational to the strategic levels [38]. Upon a thorough understanding of the interlinkages between resources, the EWF nexus can also represent a conceptual approach to improve sustainability pillars and policy-making within the three resources systems. The EWF nexus can also serve to initiate and facilitate the cfficient collaboration between resource sectors [39], which can be achieved by combining analytical tools with well thought conceptual frameworks. The three roles of the EWF nexus can be used interchangeably as they all intend to improve the performance of resource systems through promoting cross-sectoral sustainable cooperation.

3.1 Energy and Water

The EWF nexus literature has significantly focused on analysing and quantifying the convoluted interlinkages existing between the three nexus systems. Essentially, the major focus of the studies addressing the EWF nexus is related to resource management for sustainable development. Works conducted in this regard can be divided into two major categories depending on the system of interest and respective boundary. The first pool of studies, which constitutes the largest share of the nexus literature, considers the nexus or the interactions between two resources systems only. In the past, the nexus between the energy and water sectors across process and operational levels has received remarkable attention owing to its importance in driving the majority of industrial sectors leading the economy. In this respect, Wang et al. evaluated the impact of the water sector decisions on the energy strategies. A scenario analysis technique based on the input–output analysis (IOA) was adopted to investigate diverse energy-mix cases with an aim to determine pathways that could reduce inefficiencies that arise from the energy and water interdependencies [40]. The same tool was adopted by Duan and Chen in combination with the ecological network analysis (ENA) to examine the water embodied in the production of energy and international trade. Results of their study illustrated the reliance of the energy sector on the water system and the impact that this dependence induces on the latter [41]. With a more focus on the process level, Sun et al. analysed water usage in cooling towers used to dissipate the heat generated from concentrated solar power (CSP) power generation plants. The study illustrates the effectiveness in using water sprays to enhance the efficiency of CSP plants by improving the performance of natural draft dry cooling towers (NDDCT), a necessary cooling technology in arid and dry climates [42]. Aiming to enhance the efficiency of the power generation system and reduce the burden of water resources, Yasir et al. suggested a coupled power-desalination plant that captures the water vapour generated from the natural gas-fired power station, as means to recover the water used especially in regions with scarce water resources. In addition, deploying such systems can result in a significant reduction in CO_2 emissions [43]. Considering the energy to water integration, Aminfard et al. assessed the economic and environmental feasibility of using renewable energy sources such as solar and wind energy to power a reverse osmosis water desalination plant. With the two technologies resulting in reduced carbon footprints, findings have demonstrated that wind energy is more economically viable than the solar energy [44]. Similarly, Al-Obaidli et al. utilised a portfolio analysis optimisation to determine the optimal energy and water desalting configurations composed of renewable and non-renewable energy sources driving a cogeneration plant. The aim of the study was to minimise emissions associated with fuel-based energy sources while achieving a minimal levelised cost [45].

3.2 Food and Water

Interactions between the food and water sectors are also well demonstrated in the nexus literature. Generally, the focus of studies in this area is on two subsystems of the food sector, namely agriculture and international trade. For instance, Lahlou et al. developed an optimisation model that allocates treated waste water generated from diverse industries to irrigate and supply nutrients to a fodder farm. Findings of the framework asserts that using treated waste water as a source of water and fertilisers results in negative carbon emissions and reduces the stress on renewable water resources, especially in countries with characterised by a scarce water base

[46]. As for the international food trade, the food–water relationship is manifested by the concept of virtual water. It is defined as the water embodied in the food products and includes water required during production along with the water consumed throughout the supply chain [47]. Ye et al. proposed a novel multi-objective optimisation model that introduces the virtual water contained in five major crops as part of the allocation of water resources along with physical water. The model suggested generates the water distribution amongst different urban activities, including agriculture, the environment and the industry considering economic and environmental tradeoffs [48].

3.3 Energy and Food

The food sector requires water resources as the main component to grow crops and raise livestock. However, all the processes adopted to extract, treat, or desalinate water necessitate energy in the form of electricity. In addition, fertilisers, which are an essential source of nutrients to cultivate crops, are also dependent on electricity for their production [8]. This strong bond between the energy and water sector was the focus of many nexus studies. From an economic perspective, Taghizadeh-Hesary et al. investigated the impact of energy prices on food prices. Findings demonstrate a strong correlation between the prices of the two commodities, such that any volatility or inflation in oil prices can induce a significant variation in the food prices, hence threatening the economic stability of the food sector [7]. With rising concerns over climate change and the emergence of climate-positive actions to reduce the environmental impact associated with human activities, new eco-friendly technologies are deployed as part of the food–energy nexus as a form of mitigation. In fact, Haltas et al. proposed an alternative energy system based on anaerobic digestion as means to reduce the environmental burden engendered by conventional power used to produce electricity for cultivation and production of fertilisers [49]. Ghiat et al. studied a biomass-based energy system for the generation of power using a clean and efficient process analogous to the concept of bioenergy with carbon capture and storage (BECCS). A biomass-based gasification combined cycle (BIGCC) coupled with carbon capture was modelled and simulated using Aspen Plus software in efforts to study the technical performance of the system and optimise the energetic and exergetic efficiencies of the integrated system. In addition, the economic and environmental performance of the hybrid system was investigated with findings indicating negative emission levels for the proposed system. This study demonstrated an important energy–food opportunity that can be harnessed to reach negative CO_2 emissions and mitigate climate change [50, 51].

3.4   Energy, Water and Food

Adopting a holistic resource management perspective, the EWF nexus literature encompasses a wide range of studies that have expanded system boundaries to encompass more than two resource sectors. With an aim to include all the inter-linkages intrinsically relating the three systems in the assessment of the EWF nexus, Al-Ansari et al. developed a modularised EWF nexus framework that decomposes the resources sectors into a sequence of simple sub-systems enabling the characterisation of synergies and divergences amongst sectors. The proposed approach assesses the environmental performance of a food security case through investigating multiple energy, water, and food configurations. The Life Cycle Assessment (LCA) method-ology is applied to quantify the environmental performance for each scenario that meets the desired food self-sufficiency level. The work also investigated, as part of the scenario's analyses, the implementation of advanced and environmental effi-cient energy and water technologies that contribute to the reduction of the carbon and water footprints. In this regard, reverse osmosis-based desalination was consid-ered as an alternative water source to the conventional groundwater abstraction. As for the energy system, alternative configurations include photovoltaics and neutral and negative biomass power plants in the form of biomass-based integrated gasi-fication combined cycle (BIGCC) and bio-energy with carbon capture and storage (BECCS). Considering the food sector, CO_2 fertilisation was examined as an option illustrating the energy to food interaction [13, 52, 53]. These studies were followed by a techno-economic analysis investigating the performance of the proposed EWF nexus configurations developed by Govindan et al. [54]. With emphasis on the inter-actions between resource sectors, Namany et al. adopted a game theoretic approach to allocate resources optimally between the EWF nexus systems as an expansion of the work conducted by Al-Ansari et al. [50]. In addition to the process and operational level-related studies, holistic EWF nexus literature is comprised of a diversified pool of case studies that have expanded the scope of analysis to include policy implica-tions and social behaviours on the overall performance of the EWF nexus systems. For instance, Bieber et al. developed an integrated approach based on an Agent-based modelling (ABM) combined with a scenario-based optimisation that considers the influence of human and socio-economic habits on the demand on water and energy. The tailored methodology investigates the opportunity of implementing a power generation plant instead of a food production system. Economic and environmental efficiency are the two objectives desired in this model considering the operating costs and initial investments for power generation, along with the food forgone opportu-nity cost [55]. In terms of the challenges faced by the resources sectors as discussed previously, policy-makers and external entities can have a significant influence on the operations of the EWF nexus systems. In this regard, Mercure et al. emphasised the importance of considering policies and legislation in the assessment of the EWF sectors. Governance discrepancies along with social and economic variations were found to be the main stressors challenging the management of the nexus. The study suggested the alignment of science and policy-making to alleviate the existing obstacles and to enable sustainable and efficient resource management [56]. Providing a more concrete example representing a real-life illustration of the EWF nexus systems and their associated interactions, Alzaabi et al. covered the case of a smart city as a sustainable exemplar and a nest for the EWF nexus. The study has focused mainly on the role of ICT in bridging the gap between stakeholders involved in the management of the diverse sectors and organisations. Improving governance and enhancing technologies were recommended to improve the overall performance of the smart city in addition to overall EWF nexus governance [57].

Integrated analyses of the EWF nexus are necessary to achieve sustainable development and efficient resource management, which is fundamental to the fulfilment of the food security pillars [58]. Including all the subsystems, stakeholders and interlinkages is necessary to oversee the performance of the entire nexus and mitigate potential risks that might arise internally or externally. However, the major challenge in holistic nexus studies resides in the complexities that arise with the expansion of the system boundary. In fact, considering the three systems in one analysis implies diverse modelling challenges and mathematical complexities owing to the multiple stakeholders involved, the different objectives for which they aspire, in addition to the risks and uncertainties governing systems. Adopting the appropriate modelling tools and enhancing the decision-making within the EWF nexus represents a promising solution to alleviate such challenges. The nexus literature is rich with reviews that have compiled modelling and analytical tools to resolve modern resources management issues from different perspectives and disciplines based on social, economic and environmental dimensions [33, 37, 38, 59]. In the following section, the review suggested is an addition to previous literatures in the field of the EWF nexus. It alludes to the significant number of studies that have addressed the issue of food security using decision-making tools. The focus will be on solving the multiple uncertainties and risk challenges faced by food systems with the aim of a developing a resilient food system that supports security in the system.

4　EWF Nexus Modelling for Efficient Decision-Making and Resilient Food Security

The EWF nexus along with the synergistic approach that it promotes amongst the different resources systems and subsystems have been proven to be beneficial in reducing food insecurity [60]. The nexus approach enables the understanding of the multifaceted interlinkages supporting the resources systems. In addition, the collaborative systems approach that drives nexus frameworks drive the shift in policy-making and legislation from the conventional narrow lens, towards a more integrated reasoning harnessing all sectors for optimal nexus performance.

The literature has demonstrated the benefits of EWF nexus modelling tools in improving the efficiency of food sectors through the mutual maximisation of benefits and gains amongst sectors, the minimisation of uncooperative competitions, which usually lead to collective losses and inefficiencies, in addition to the mitigation of the internal factors that threaten interactions between sectors and the overall performance of the nexus [61]. The improvement that modelling and analytical tools bring to the EWF nexus and food systems has proven its effectiveness across the major pillars of food security. For instance, enhancing the crop production [62] and storage of agricultural products [63], in addition to reducing food waste in the food system supply chain [64], have contributed to the enhancement of the food availability pillar. This was achieved by means of optimisation models and analytical tools that ensure the communication between the three resources systems. Similarly, models targeting the adjustment of the dietary needs while considering the water and energy consumption considering economic and environmental concerns have contributed to the food utilisation component of the food security [65, 66]. As for food accessibility, which accounts for household income and food prices; a holistic EWF nexus system approach might support enhanced food affordability and grant a wider access to food products at more affordable costs [67, 68].

In order to ensure the stability of the three pillars of food security, internal and external risks and uncertainties must be factored into EWF nexus modelling and mitigated accordingly. In fact, internal risks associated with each sector's operations, and external risks inflicted by the surrounding environment should be considered while making decisions and planning for sustainable, resilient, and long-lasting food security. As such, the following section will highlight the tools and models, which can be applied to understand risks and uncertainties in volatile environments, and that can form the basis of control and mitigation to sustain the performance of the EWF nexus as a means for food security. First, simulation models used to virtually represent resource systems will be discussed. Subsequently, risk quantification and prediction techniques will be presented to realistically depict real-life uncertainties. Finally, optimisation-based models which are the most used technique to enhance the performance of systems will be highlighted. The three proposed modelling categories suggest a realistic logic that can be followed to enhance food systems, while accounting for the uncertainty and the dynamic in the systems. The three categories can be used jointly (see Fig. 1) or independently to solve diverse issues altering the food security target.

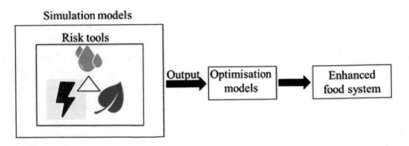

Fig. 1 High-level illustration of the decision-making models for resilient food systems

4.1 Simulation Models

The food sector, and the EWF nexus framework to which it belongs, is comprised of a set of systems and sub-systems characterised by a range of operational complexities owing to the diverse processes they require to function. Furthermore, the environment surrounding these systems is a nest of various risks emerging due to societal, economic, environmental and geopolitical circumstances that can induce several volatilities and instabilities [61]. Under these pressures exerted on the food sector and its associated resource systems (water and energy), which could result in unpredictable events that occur at any given point in time; investing in food technologies and related projects without prior risk assessment and control methodologies might be adventurous and could engender undesirable results, such as financial losses and environmental damages. Simulation models mitigate such systemic failures by providing a reality-like representation of systems and allow for future planning considering exogenous events. Simulation modelling is defined as the procedure used to virtually represent a real-life problem such us a systemic process in order to predict its behaviours. It allows for the identification of strengths and bottlenecks in the system as a means for performance enhancement and contribute towards avoiding the snowball effect [69]. Simulation models are used to assist policy-makers and stakeholders in any discipline in the understanding of the output of their decisions and strategies prior to real-life implementation. It also enables the dynamic interactions and flexible communication between different stakeholders and sectors which result in maximising payoffs and shared mitigation of uncertainties and risks. Furthermore, simulations models and software, contrary to the physical modelling that are time consuming and expensive permit the possibility to modify and upgrade the experiment during the running time of the simulations providing more control over the system of study. In the context of food security, many simulation modelling techniques have been used to enhance the performance of the food sector. In the following sections, three widely used simulation methods will be discussed to illustrate the effectiveness of simulation techniques in providing an opportunity for risk mitigation as part of the food security target.

4.1.1 Discrete Event Simulation

Discrete event simulation (DES) is an extensively adopted simulation technique used in instances where the process or operation of a system is structured as a sequence of events [70]. In DES, the state variables vary at some specific points in time representing the events, incidentally, any change in the system modelled using DES occurs only in the advent of an event, and no variations are witnessed over the entire modelling life-time. State variables, representing the independent variables, are characterised with particular exclusive attributes defining their parameters and functionalities. Information associated with each attribute is subject to variation across the time of the simulation if any change occurs to the status of parameters. Another central component that constitutes a DES is the resources constituting the information that are stored and updated across the running time of the simulation. In DES, state variables are passive entities that can only be controlled through their attributes and associated resources. In the context of decision-making the DES enables the testing of "what if" scenarios, allowing decision-makers a flexible, efficient and prior assessment of alternatives as means to avoid the risk of potential future failures in the system due uninformed decisions. In addition, DES is usually applied in systems requiring complex process modelling and where decisions are often intricate [71]. The food system, one of these complex systems, is a sector where DES has been extensively used to prevent risks and respond to unexpected events [72].

DES appears frequently in EWF nexus literature as part of efforts to attain food security targets. Studies in this regard have covered the three essential pillars of food security with an aim of enhancing food system resilience and stability. Focusing on crop cultivation, a building block of "food availability", Van't Ooster et al. adopted a DES model to simulate the crops' cultivation inside greenhouses. The model used is the Greenhouse Work Simulation (GWorkS), which is used to improve labour conditions for sweet pepper harvesting. The developed dynamic model is used to formulate technical and economic restrictions that would later help design an innovative labour conditions for an enhanced cultivation system [73]. The GWorkS model used was originally developed by the same authors, Van't Ooster et al., to increase labour and machine efficiencies in a greenhouse based on mobile rose systems. The purpose of the model is to test operational scenarios related to labour work and machinery prior to the implementation. The tailored DES is a generic tool that can accommodate any greenhouse crop's harvesting process [74]. Considering another system from the agricultural sector, Gittins et al. proposed a hybrid framework based on a DES model to manage livestock farms, the model utilises a combination of simulations and surveys to model strategies of the farmers. This work comes as a response to the volatile social and political conditions that have followed the decision for Brexit decision from the UK. Findings of the research present diverse growth scenarios that capture the opinions of farmers along with simulated empirical evidences that suggest the potential technology adjustments that could improve the livestock farming sector [75]. Supply chain and logistics of food products is also an area where DSE have contributed to its improvement. For instance, Nilsson used the DES to analyse the logistics system of reed canary grass and straw, which are plants usually used as animal forage. However, in this example, the fodder was used as a source of heat for a water system. The DES model developed provides the optimal mix of technologies, machines, fodder quantities and capacity of storage that ensure the supply of the forage from the field to the heating plant [76]. In addressing food quality, Van der Vorst et al. proposed an integrated approach based on a DSE model that enhances food supply chains and ensures that demands are satisfied while maintaining food quality and sustainability principles [77]. Lopes et al. focused on the transportation aspect of food supply chains through developing a DES model that simulates the transportation network of soybeans export system. Considering the total costs of the process, results of this study provided decision-making insights related to the strategic and operational strategies that can be undertaken in the selected ports to enhance the supply chain system [78].

4.1.2 System Dynamics

EWF nexus sectors are complex systems involving multifaceted interactions that require advanced decision-making process. Furthermore, such systems are subject to multiple changes over time due to the variations in their subsystems and surrounding environments. While the DES can accommodate the changes in the system by simulating disruptive events and proposing enhancement scenarios and strategies, it can only simulate the response of the system with events occurring in some specific instances in time. Resource systems, including the food sector, require continuous managerial and policy involvements governed by unending disturbances that should be captured continuously over time. In order to accommodate for such aspects in the system, the System Dynamic (SD) approach is a simulation technique that allows for the continuous decision-making. In fact, the focus of SD goes beyond the analysis of individual events, instead, it analyses the dynamic rules and patterns governing them. In an SD model, the emphasis is not on the events and the discrete decisions they engender, however, the major focus is on the founding policies constructing such decisions [79, 80]. In other words, events along with their derivative decisions are considered a superficial result of a convoluted interaction between the system's elements. It is also worth mentioning that in dynamic problems where DS is deployed, variability consists of a causality relationship represented by feedback loops that constitute the interactions between variables/entities, and which are reflected in the behaviour of the entire system. Typically, SD models are used to simulate challenging and large-scale systems governed by socio-economic variabilities [81]. The EWF nexus resource systems are a suitable case where SD modelling can be applied, as the latter can be used to capture the dynamics amongst and within the three systems as means to improve managerial decisions and hence the system performance [82]. In this regard, SD was heavily used to enhance the food security issue through investigating the dynamics within the food sector's systems. Tsolakis and Srai developed a holistic methodology based on a SD model to study the impact of smallholder farming on sustainability issues and particularly on food security. The framework considers the self-sufficiency as a metric for food security. Findings of the scenarios generated by the model indicate that improving the policy and governance in smallholding farming, in addition to the fostering of short supply chains of food products could lead to a further increase in self-sufficiency leading to an improved food security [83]. With emerging concerns for increasing environmental impacts associated with the utilisation of resource systems; sustainability and EWF nexus literature have shifted the attention to technologies and inventions that would reduce stresses on the environment. In this regard, Martínez-Jaramillo et al. investigated the impact of investing in biofuel production as an alternative fuel to reduce emissions caused by the transportation sector. A SD model was used to assess the influence of such decisions on food security. The model simulates the relationship between biofuel production, food production and raising of livestock. Outcomes indicate a negative effect on food security, as investing in biofuels reduce lands available for agriculture causing food shortages and therefore causing a surge in food prices [84]. DS modelling can be also used in assessing social concerns associated with food security. For instance, Galli et al. investigated the relationship between the disproportionate food security across the globe due to poverty in low-income countries, and excessive food waste in wealthier nations and developed regions. In order to alleviate these two social issues, a holistic framework based on a DS model that depicts the behaviour of diverse stakeholders involved in the food sector stream, including consumers, retailers, farmers and caterers, is used to map the links between waste generation, food redistribution for social causes and food poverty [85].

4.1.3 Agent-Based Modelling

Considering an agriculture farm that involves a fodder production entity and livestock farming, as an example, making a decision to grow a specific fodder type in a particular land under certain climatic conditions would require a deep and comprehensive understanding of the requirement of the crop in terms of water consumption and nutrient intake. Considering that SD is a top-down approach that analyses a system using an aggregated perspective would not be effective in capturing the individual characteristics of each crop. Instead, it would provide an overview on the impact of cultivating that fodder on the raising of livestock, as a different component of the analysed system. SD is efficient in depicting the relationships between diverse elements of the system; however, it does not clearly depict the inner functionalities for each unit of the system [86]. Agent-based modelling (ABM) is an alternative approach that can simulate the system while intricately capturing the functionalities of its smallest components. The ABM follows a bottom-up approach that allows the communication between independents entities known as agents. It is characterised by modularity in simulation, which enables the aggregation of complex and large-scale problems into a set of sub-systems that are studied individually and are then aggregated to form a complete and dynamic solution [87].

ABM is usually applied to depict the interactions between independent agents coexisting in an instable environment. It simulates the interactions amongst agents in addition to the interactions between agents and their surrounding environment. Agents in an ABM are independent entities represented by some exclusive attributes and behaviours. The interactions between agents and the responses to the environment are governed by a set of behavioural rules established by agents or inflicted by external influences [88]. ABM offers a flexible and dynamic ground to solve complex system involving multiple stakeholders with divergent or convergent targets by allowing them to set their rules and interact freely to achieve their goals. This dynamic property in the ABMs is very efficient to direct decision-making in complex and multi-disciplinary problems, principally those comprising of multiple levels and stakeholders, such as EWF nexus resource systems. As such, food security, which is one of the most stressing issues tackled by the EWF nexus literature, encompasses numerous studies adopting ABMs. Wens et al. developed a risk adaptation framework to simulate drought adaptation behaviours in agricultural sector. The purpose of the study was to determine the impact of adopting adaption measures on the drought risk profile in a semi-arid environment [89]. Considering food security from a social perspective, Joyita developed a framework that fosters food donation through enhancing the collaboration within lucrative entities known as food pantries. The ABM-based approach simulates the supply and demand profiles on food and the network facilitating the flow of food between different pantries, with an aim to enhance collaboration amongst them and alleviate food insecurity [90]. ABM was also used to enhance the economic and environmental efficiency of food systems by assessing diverse production and supply scenarios. Namany et al. proposed a flexible decision-making framework that predicts strategies of a food sector comprised of local production and international trade systems. Three scenarios were developed to investigate the food supply profile under water restrictions constraint and deployment of forward contracts opportunity. The methodology developed was illustrated by tomato crop, yet it can accommodate diverse crop type. It fosters the importance of considering the virtual water concept as part of the water resources in arid regions, and it serves as a decision-making guideline to improve the performance of the food sector while respecting resources scarcities and economic limitations [91].

4.2 Risk and Uncertainty Assessment and Prediction

The world today is subjected to the increasing emergence of social, economic, and environmental challenges. Climate change, political instability, natural disasters, and demographic growth are only but a few examples of the myriad of turbulences that can threaten the stability and prosperity of mankind. However, the demand for products and resources to maintain the current operations in all sectors necessitates the need to overcome instabilities and cope with current and unforeseen uncertainties either through mitigation or adaptation strategies. In order to develop mitigation plans, risks should be first identified and quantified to capture the influence they may exert on diverse systems.

In the previous Sect. 4.1, simulation methods were explained, and their importance in depicting real-life problems was highlighted. However, in order to have a comprehensive and realistic representation of current issues, risks and uncertainties should be embedded within existing models to accurately mimic systems interactions and dynamics since real problems are rarely deterministic, and mostly involve stochasticity and randomness. There exists a wide variety of risk quantification models and techniques that are either used independently or jointly with the aforementioned simulation models. In the context of resource management, the focus in this chapter will be mainly on common techniques used to quantify and predict risks as part of the food insecurity alleviation goal and building a resilient food sector.

Risk assessment is broadly defined as the steps followed to identify the probability of potential losses, through the analysis of vulnerabilities and hazards governing the surrounding environment of a system, livelihoods or people [92]. The aim of risk assessment and prediction within resource management to determine and predict the possible risks and uncertainties that can cause systemic failures and reduce the system's adaptive capacity. In addition, it can enable the effective evaluation of existing control actions and re-adjust according to the predicted risks. Any failure to account for risks in one of the resource sectors can cause a cascade failure due to the inherent interlinkages between the resource systems. One example of cascade failure can be illustrated with the energy to food interaction. For instance, volatility in gas prices, which is considered a risk in the energy sector, can influence the energy production, which can alter the supply of the electricity, a critical component for irrigation technologies, production of fertilisers and to support farms [93]. This sequence of disturbances can reduce crop production and engender food shortages. To avoid such undesirable results, considering volatility in prices and uncertainties while making decisions can reduce potential losses. One of the tools commonly used to capture risks associated with the food sector and its associated energy and water sectors is the Monte Carlo simulation [94]. Liu et al. utilised the Monte Carlo simulation with a stochastic programing model to assess the impact of water inflow uncertainty on the planning of irrigation water and crops. The purpose of the study is to alleviate risk of food shortages due to uncertain water availability [95]. Monte Carlo also supports the assessment of financial risks associated with food-related investments and projects. In this regard, Kadigi et al. adopted a Monte Carlo simulation that incorporates the risk preferences for decision-makers in the deployment of enhanced rice farming practices and technologies. To this end, the economic feasibility of each project is assessed and used as a criterion to select the most economically profitable in terms of rice productivity [96]. In addition to simulation, data-driven models are often used to account for risks and uncertainty in modelling resource systems. Machine learning is one category of data-driven models that has shown its effectiveness in accurately modelling large-scale intricate problems. While simulation tools begin with unknown input variables, representing the source of uncertainty, the model that transforms inputs into outputs, in the form of mathematical relationships and equations, is known in advance. Alternatively, machine learning inputs are known in advance while the model is unknown. In this case, historical data is used to draw the relationship between the inputs and outputs, making the model the only source of uncertainty. In other words, simulation models use the knowledge of experts about models to predict outputs, while machine learning design the models to predict results [97]. Using machine learning for food security purposes, Kumar et al. used a machine learning algorithm to predict crop yields. The purpose of the model is to determine the set crops with the highest productivity or yield while considering a range of factors, such as weather conditions, crop type, soil type and water density [98]. Similarly, Kuwata and Shibasaki used a deep learning approach, a sub-category of machine learning to predict crop yields, presenting a more accurate version of the previous crop yield model by reducing reliance on input data [99]. Another application of machine learning in agriculture is presented by Bagheri et al.

who assessed the impact of environmental contaminants' accumulation on crops quality as a major concern in food security. The study utilised a neural network to predict the contaminants concentration factor in plants [100].

4.3 Optimisation Models

While simulation models are capable of describing real-life systems with all their interacting stakeholders and entities, risk quantification and assessment techniques offer an opportunity to account for all uncertainties that disturb systems and reduce their efficiencies. Alternatively, optimisation models are used to enhance the overall performance of systems. They are usually used as a complementary tool to the simulation models, such that the output from the simulation is used as an input in the optimisation [101]. Optimisation is a widely used technique in decision-making as it can accommodate all decision levels, counting the strategic, tactical and operational offering a promising opportunity to solve problems in multifaceted food systems [38, 102]. Optimisation is primarily used to improve the performance of any system/sector by enabling the selection of the best solution that maximises benefits and minimises losses. It is a mathematical representation of a problem requiring the selection of an alternative amongst a set of choices given a set of constraints [103]. Throughout the history of its application, optimisation models have demonstrated their effectiveness in addressing modern resource management issues through holistically tackling their challenges. In the context of the EWF nexus systems and the food security targets, optimisation has assisted in alleviating the multi-dimensional challenges of food systems. It was used to balance the diverse objectives by means of multiobjective optimisation. For instance, Namany et al. developed a framework based on a multi-objective optimisation model to determine the optimal energy and water mix to maximise food self-sufficiency, while minimising economic and environmental costs [104]. Similarly, Al-Thani et al. adopted the same method to maximise both the self-sufficiency for some specific food categories along with their nutritional value, while considering some predetermined water and energy amounts [65]. Optimisation techniques are also widely used in instances where the system is undergoing uncertainties. For short-term risks, Karan et al. suggested a stochastic optimisation model that minimises the capital and operating costs of a greenhouse considering stochasticity in energy supply [105]. For long-term risks, Beh et al. proposed an approach based on robust optimisation to account for deep uncertainty affecting a system for water supply. Strong uncertainty is due to variability in climatic conditions and changes in population dynamics [106]. As mentioned previously, resources and food systems involve multiple sub-systems and entities governed by multiple stakeholders. Quite often, stakeholders have different visions and goals that might create unhealthy competition and negatively influence the performance of systems. To transform the emerging competition into a collaborative cooperation, cooperative game theoretic approaches are used as part of optimisation models. Namany et al. proposed a linear programming model with a Stackelberg game theoretic approach

Table 1 Summary table of common optimisation models (modified from Namany et al. [38])

Model	Definition
Multi-objective optimisation	A category of multi-criteria decision-making (MCDM) adopted for cases with different convergent or divergent objectives. It is used to alleviate the multi-objective challenges of resources systems [108]
Robust optimisation	Robust optimisation is used to optimise problems with extreme uncertainty and evaluate worst-case scenarios. It is used to alleviate the multi-uncertainty challenge of resources systems [109]
Stochastic optimisation	Stochastic optimisation is adopted in problems involving randomness and short-term uncertainties. It is used to alleviate the multi-uncertainty challenge of resources systems [110]

to account for competition between the energy sectors in producing electricity for a food production system. The purpose of the methodology is to minimise the cost of the EWF nexus system responsible for producing food while mitigating the competition between two power generation systems [93]. Kashid proposed an optimisation model based on game theory to consider the competition between farmers that aim to maximise their profits from crop cultivation. The model also provides an opportunity to investigate alternative strategies that use the minimum amount of resources to produce larger quantities [107]. Table 1 presents a concise summary of the commonly used optimisation techniques in addressing food security issues.

5 Conclusion

Food insecurity is a persistent global challenge and is part of the sustainable development agenda. The representative food system is multifaceted in nature involving multiple systems across various supply chains and is exposed to diverse anthropogenic and environmental uncertainties. Overcoming food insecurity necessitates the understanding of the different interlinkages that exist between the energy, water and food sectors. As such, the EWF nexus system approach enables the characterisation and understanding of such relationships and can support food security targets. Modelling EWF nexus systems can be supported using diverse decision-making tools that can provide a real-life representation of food systems by means of simulation models, while accounting for risks associated with volatile and uncertain environments using risk quantification tools, and finally generating optimal and enhanced solution using optimisation models.

To improve the ability of the EWF nexus systems approach in solving resource management issues and particularly food security, future research studies should focus on developing dynamic decision frameworks capable of accommodating the ever-changing environments and volatile economies. This should be coupled with the adoption of consolidative perspectives that are not focused on improving food security, solely, but targeting the enhancement of water and energy securities, being the building-blocks of food provision and supply chains. In addition, more emphasis should be granted to predictive and simulation models as means to assist policy-makers in taking efficient decisions prior to the implementation stage. This would serve as a preventive measure against potential systemic failures that would occur in the future and which can hinder the achievement of resources security.

References

1. Zhou H (2009) Population growth and industrialization. Econ Inq. https://doi.org/10.1111/j.1465-7295.2008.00151.x
2. United Nations (2015) World population prospects: the 2015 revision. https://doi.org/10.1007/s13398-014-0173-7.2
3. UNESCO. Global water resources under increasing pressure from rapidly growing demands and climate change, according to new UN world water development report
4. World Food Summit (1996) Rome Declaration on world food security and World Food Summit plan of action. World Food Summit
5. Ericksen PJ, Ingram JSI, Liverman DM (2009) Food security & global environmental change: emerging challenges. Environ Sci Policy 12:5. https://doi.org/10.1109/TR.2012.2194173
6. FAO (2017) Water for sustainable food and agriculture
7. Taghizadeh-Hesary F, Rasoulinezhad E, Yoshino N (2019) Energy and food security: linkages through price volatility. Energy Policy. https://doi.org/10.1016/j.enpol.2018.12.043
8. Garcia DJ, You FQ (2016) The water-energy-food nexus and process systems engineering: a new focus. Comput Chem Eng 91:49–67. https://doi.org/10.1016/j.compchemeng.2016.03.003
9. Shine P, Scully T, Murphy MD (2018) Predicting annual electricity consumption on Irish pasture-based dairy farms using a support vector machine model
10. Ortiz XA et al (2015) Evaluation of conductive cooling of lactating dairy cows under controlled environmental conditions. J Dairy Sci. https://doi.org/10.3168/jds.2014-8583
11. Sur A, Sah RP, Pandya S (2020) Milk storage system for remote areas using solar thermal energy and adsorption cooling. Mater Today Proc. https://doi.org/10.1016/j.matpr.2020.05.170
12. Lamidi RO, Jiang L, Pathare PB, Wang YD, Roskilly AP (2019) Recent advances in sustainable drying of agricultural produce: a review. Appl Energy. https://doi.org/10.1016/j.apenergy.2018.10.044
13. Al-Ansari T, Korre A, Nie Z, Shah N (2017) Integration of greenhouse gas control technologies within the energy, water and food nexus to enhance the environmental performance of food production systems. J Clean Prod 162:1592–1606. https://doi.org/10.1016/j.jclepro.2017.06.097
14. Hua E, Wang X, Engel BA, Sun S, Wang Y (2020) The competitive relationship between food and energy production for water in China. J Clean Prod 247. https://doi.org/10.1016/j.jclepro.2019.119103
15. Ignaciuk A, Vöhringer F, Ruijs A, van Ierland EC (2006) Competition between biomass and food production in the presence of energy policies: a partial equilibrium analysis. Energy Policy. https://doi.org/10.1016/j.enpol.2004.09.010
16. Best R, Burke PJ, Jotzo F (2020) Carbon pricing efficacy: cross-country evidence. Environ Resour Econ. https://doi.org/10.1007/s10640-020-00436-x
17. AlNouss A, Namany S, McKay G, Al-Ansari T (2019) Applying a sustainability metric in energy, water and food nexus applications; a biomass utilization case study to improve investment decisions. In: Computer aided chemical engineering

18. Konadu DD et al (2015) Land use implications of future energy system trajectories—the case of the UK 2050 carbon plan. Energy Policy. https://doi.org/10.1016/j.enpol.2015.07.008

19. FAO (2009) How to feed the world in 2050. Insights from an expert meet. FAO. https://doi.org/10.1111/j.1728-4457.2009.00312.x

20. Kopittke PM, Menzies NW, Wang P, McKenna BA, Lombi E (2019) Soil and the intensification of agriculture for global food security. Environ Int 132:105078. https://doi.org/10.1016/j.envint.2019.105078

21. Paul BK et al (2020) Reducing agro-environmental trade-offs through sustainable livestock intensification across smallholder systems in Northern Tanzania. Int J Agric Sustain. https://doi.org/10.1080/14735903.2019.1695348

22. Food and Agriculture Organisation (2010) Facts: the state of the world's land and water resources, p 2010

23. Vining KC (1990) Effects of weather on agricultural crops and livestock: an overview. Int J Environ Stud. https://doi.org/10.1080/00207239008710581

24. Mousa H et al (2020) The role of urban farming in revitalizing cities for climate change adaptation and attaining sustainable development: case of the City of Conegliano, Italy

25. Haji M, Govindan R, Al-Ansari T (2020) Novel approaches for geospatial risk analytics in the energy-water-food nexus using an EWF nexus node. Comput Chem Eng. https://doi.org/10.1016/j.compchemeng.2020.106936

26. Aggarwal A, Upadhyay R (2013) Shelter management for alleviation of heat stress in cows and buffaloes. In: Heat stress and animal productivity

27. FAO (2017) Drought characteristics and management in Central Asia and Turkey

28. United States Department of Agriculture (2017) Rail disruptions following Hurricane Harvey nearly halt grain deliveries to Texas Gulf ports. https://www.ers.usda.gov/data-products/chart-gallery/gallery/chart-detail/?chartId=85259

29. Chen K, Zhan Y, Zhang Y, Fan S (2020) The impacts of COVID-19 on global food security and the coping strategy. China Rural Econ

30. Stephens EC, Martin G, van Wijk M, Timsina J, Snow V (2020) Editorial: impacts of COVID-19 on agricultural and food systems worldwide and on progress to the sustainable development goals. Agric Syst. https://doi.org/10.1016/j.agsy.2020.102873

31. FAO (2014) The water-energy-food nexus—a new approach in support of food security and sustainable agriculture. Food and Agriculture Organization of the United Nations. https://doi.org/10.1039/C4EW90001D

32. Hoff H (2011) Understanding the nexus. Background paper for the Bonn 2011 conference: the water, energy and food security nexus

33. Mannan M, Al-Ansari T, Mackey HR, Al-Ghamdi SG (2018) Quantifying the energy, water and food nexus: a review of the latest developments based on life-cycle assessment. J Clean Prod. https://doi.org/10.1016/j.jclepro.2018.05.050

34. IEA (2011) Energy for all: financing access for the poor. In: World energy outlook 2011

35. Sadoff C, Grey D, Borgomeo E (2020) Water security. In: Oxford research encyclopedia of environmental science, Oxford University Press

36. Maass M (2017) Integrating food-water-energy research through a socio-ecosystem approach. Front Environ Sci. https://doi.org/10.3389/fenvs.2017.00048

37. Albrecht TR, Crootof A, Scott CA (2018) The water-energy-food nexus: a systematic review of methods for nexus assessment. Environ Res Lett. https://doi.org/10.1088/1748-9326/aaa9c6

38. Namany S, Al-Ansari T, Govindan R (2019) Sustainable energy, water and food nexus systems: a focused review of decision-making tools for efficient resource management and governance. J Clean Prod 225:610–626. https://doi.org/10.1016/j.jclepro.2019.03.304

39. Keskinen M, Guillaume JHA, Kattelus M, Porkka M, Räsänen TA, Varis O (2016) The water-energy-food nexus and the transboundary context: insights from large Asian rivers. Water (Switzerland). https://doi.org/10.3390/w8050193

40. Wang S, Fath B, Chen B (2019) Energy–water nexus under energy mix scenarios using input–output and ecological network analyses. Appl Energy. https://doi.org/10.1016/j.apenergy.2018.10.056

41. Duan C, Chen B (2017) Energy–water nexus of international energy trade of China. Appl Energy. https://doi.org/10.1016/j.apenergy.2016.05.139
42. Sun Y, Guan Z, Gurgenci H, Wang J, Dong P, Hooman K (2019) Spray cooling system design and optimization for cooling performance enhancement of natural draft dry cooling tower in concentrated solar power plants. Energy. https://doi.org/10.1016/j.energy.2018.11.111
43. Yasir AT, Eljack F, Kazi MK (2020) Synthesis of water capture technologies for gas fired power plants in Qatar. Chem Eng Res Des 154:171–181. https://doi.org/10.1016/j.cherd.2019.12.013
44. Aminfard S, Davidson FT, Webber ME (2019) Multi-layered spatial methodology for assessing the technical and economic viability of using renewable energy to power brackish groundwater desalination. Desalination. https://doi.org/10.1016/j.desal.2018.10.014
45. Al-Obaidli H, Namany S, Govindan R, Al-Ansari T (2019) System-level optimisation of combined power and desalting plants. In: Computer aided chemical engineering
46. Lahlou F-Z, Mackey HR, McKay G, Onwusogh U, Al-Ansari T (2020) Water planning framework for alfalfa fields using treated wastewater fertigation in Qatar: an energy-water-food nexus approach. Comput Chem Eng. https://doi.org/10.1016/j.compchemeng.2020.106999
47. Guan D, Hubacek K (2007) Assessment of regional trade and virtual water flows in China. Ecol Econ. https://doi.org/10.1016/j.ecolecon.2006.02.022
48. Ye Q et al (2018) Optimal allocation of physical water resources integrated with virtual water trade in water scarce regions: a case study for Beijing, China. Water Res. https://doi.org/10.1016/j.watres.2017.11.036
49. Haltas I, Suckling J, Soutar I, Druckman A, Varga L (2017) Anaerobic digestion: a prime solution for water, energy and food nexus challenges. Energy Procedia 123:22–29. https://doi.org/10.1016/j.egypro.2017.07.280
50. Ghiat I, AlNouss A, McKay G, Al-Ansari T (2020) Biomass-based integrated gasification combined cycle with post-combustion CO_2 recovery by potassium carbonate: techno-economic and environmental analysis. Comput Chem Eng 135:106758. https://doi.org/10.1016/j.compchemeng.2020.106758
51. Ghiat I, AlNouss A, McKay G, Al-Ansari T (2020) Modelling and simulation of a biomass-based integrated gasification combined cycle with carbon capture: comparison between monoethanolamine and potassium carbonate. https://doi.org/10.1088/1755-1315/463/1/012019
52. Al-Ansari T, Korre A, Nie Z, Shah N (2015) Development of a life cycle assessment tool for the assessment of food production systems within the energy, water and food nexus. Sustain Prod Consum 2:52–66. https://doi.org/10.1016/j.spc.2015.07.005
53. Al-Ansari T, Govindan R, Korre A, Nie Z, Shah N (2018) An energy, water and food nexus approach aiming to enhance food production systems through CO_2 fertilization. Comput Aided Chem Eng
54. Govindan R, Al-Ansari T, Korre A, Shah N (2018) Assessment of technology portfolios with enhanced economic and environmental performance for the energy, water and food nexus. Comput Aided Chem Eng 43:537–542
55. Bieber N et al (2018) Sustainable planning of the energy-water-food nexus using decision making tools. Energy Policy 113:584–607. https://doi.org/10.1016/j.enpol.2017.11.037
56. Mercure JF et al (2019) System complexity and policy integration challenges: the Brazilian energy-water-food nexus. Renew Sustain Energy Rev. https://doi.org/10.1016/j.rser.2019.01.045
57. Alzaabi M, Rizk Z, Mezher T (2019) Linking smart cities concept to energy-water-food nexus: the case of Masdar City in Abu Dhabi, UAE. In: Smart cities in the Gulf
58. Gil JDB, Reidsma P, Giller K, Todman L, Whitmore A, van Ittersum M (2019) Sustainable development goal 2: improved targets and indicators for agriculture and food security. Ambio. https://doi.org/10.1007/s13280-018-1101-4
59. Lixiao Z, Pengpeng Z, Yan H, Shoujuan T, Gengyuan I (2019) Urban food-energy-water (few) nexus: conceptual frameworks and prospects. Shengtai Xuebao/Acta Ecol Sin. https://doi.org/10.5846/stxb201809081926

60. Rasul G, Sharma B (2016) The nexus approach to water–energy–food security: an option for adaptation to climate change. Clim Policy. https://doi.org/10.1080/14693062.2015.1029865
61. Govindan R, Al-Ansari T (2019) Computational decision framework for enhancing resilience of the energy, water and food nexus in risky environments. Renew Sustain Energy Rev. https://doi.org/10.1016/j.rser.2019.06.015
62. Alburquerque JA et al (2013) Enhanced wheat yield by biochar addition under different mineral fertilization levels. Agron Sustain Dev. https://doi.org/10.1007/s13593-012-0128-3
63. Irani Z et al (2017) Managing food security through food waste and loss: small data to big data. Comput Oper Res. https://doi.org/10.1016/j.cor.2017.10.007
64. Mogale DG, Kumar M, Kumar SK, Tiwari MK (2018) Grain silo location-allocation problem with dwell time for optimization of food grain supply chain network. Transp Res Part E Logist Transp Rev 111:40–69. https://doi.org/10.1016/j.tre.2018.01.004
65. Al-Thani NA, Govindan R, Al-Ansari T (2020) Maximising nutritional benefits within the energy, water and food nexus. J Clean Prod 266:121877. https://doi.org/10.1016/j.jclepro.2020.121877
66. Mortada S, Abou Najm M, Yassine A, El Fadel M, Alamiddine I (2018) Towards sustainable water-food nexus: an optimization approach. J Clean Prod 178:408–418. https://doi.org/10.1016/j.jclepro.2018.01.020
67. Vats G (2019) A nexus approach to energy, water, and food security policy making in India. University of Technology Sydney
68. Rajikan R, Shin LH, Hamid NIA, Elias SM (2019) Food insecurity, quality of life, and diet optimization of low income university students in Selangor, Malaysia. J Gizi Dan Pangan. https://doi.org/10.25182/jgp.2019.14.3.107-116
69. Altiok T, Melamed B (2007) Simulation modeling and analysis with ARENA
70. Katsaliaki K, Mustafee N (2011) Applications of simulation within the healthcare context. J Oper Res Soc. https://doi.org/10.1057/jors.2010.20
71. Dagpunar JS, Fishman GS (1980) Principles of discrete event simulation. J Oper Res Soc. https://doi.org/10.2307/3009344
72. Garbolino E, Chery JP, Guarnieri F (2016) A simplified approach to risk assessment based on system dynamics: an industrial case study. Risk Anal. https://doi.org/10.1111/risa.12534
73. Van't Ooster A, Aantjes GWJ, Melamed Z (2017) Discrete event simulation of crop operations in sweet pepper in support of work method innovation. https://doi.org/10.17660/ActaHortic.2017.1154.19
74. Van't Ooster A, Bontsema J, Van Henten EJ, Hemming S (2012) GWorkS—a discrete event simulation model on crop handling processes in a mobile rose cultivation system. Biosyst Eng. https://doi.org/10.1016/j.biosystemseng.2012.03.004
75. Gittins P, McElwee G, Tipi N (2020) Discrete event simulation in livestock management. J Rural Stud 78:387–398. https://doi.org/10.1016/j.jrurstud.2020.06.039
76. Nilsson D (2001) Discrete event simulation as a performance analysis tool in agricultural logistics systems. IFAC Proc 34(26):19–24. https://doi.org/10.1016/s1474-6670(17)33626-1
77. van der Vorst JGAJ, van der Zee D-J, Tromp S-O (2010) Simulation modelling for food supply chain redesign
78. dos Santos Lopes H, da Silva Lima R, Leal F, de Carvalho Nelson A (2017) Scenario analysis of Brazilian soybean exports via discrete event simulation applied to soybean transportation: the case of Mato Grosso State. Res Transp Bus Manag. https://doi.org/10.1016/j.rtbm.2017.09.002
79. Forrester JW (1968) Industrial dynamics—after the first decade. Manage Sci. https://doi.org/10.1287/mnsc.14.7.398
80. Forrester JW (1997) Industrial dynamics. J Oper Res Soc. https://doi.org/10.1057/palgrave.jors.2600946
81. Shen Q, Chen Q, Tang B-S, Yeung S, Hu Y, Cheung G (2009) A system dynamics model for the sustainable land use planning and development. Habitat Int. https://doi.org/10.1016/j.habitatint.2008.02.004

82. Saysel AK, Barlas Y, Yenigün O (2002) Environmental sustainability in an agricultural development project: a system dynamics approach. J Environ Manage. https://doi.org/10.1006/jema.2001.0488

83. Tsolakis N, Srai JS (2017) A system dynamics approach to food security through smallholder farming in the UK. Chem Eng Trans. https://doi.org/10.3303/CET1757338

84. Martínez-Jaramillo JE, Arango-Aramburo S, Giraldo-Ramírez DP (2019) The effects of biofuels on food security: a system dynamics approach for the Colombian case. Sustain Energy Technol Assess. https://doi.org/10.1016/j.seta.2019.05.009

85. Galli F, Cavicchi A, Brunori G (2019) Food waste reduction and food poverty alleviation: a system dynamics conceptual model. Agric Hum Values. https://doi.org/10.1007/s10460-019-09919-0

86. Martin R, Schlüter M (2015) Combining system dynamics and agent-based modeling to analyze social-ecological interactions—an example from modeling restoration of a shallow lake. Front Environ Sci. https://doi.org/10.3389/fenvs.2015.00066

87. Barbati M, Bruno G, Genovese A (2012) Applications of agent-based models for optimization problems: a literature review. Expert Syst Appl 39(5):6020–6028. https://doi.org/10.1016/j.eswa.2011.12.015

88. Macal CM, North MJ (2006) Tutorial on agent-based modeling and simulation part 2: how to model with agents. In: Proceedings—winter simulation conference, pp 73–83. https://doi.org/10.1109/WSC.2006.323040

89. Wens M et al (2020) Simulating small-scale agricultural adaptation decisions in response to drought risk: an empirical agent-based model for semi-arid Kenya. Front Water. https://doi.org/10.3389/frwa.2020.00015

90. Joyita M (2019) Analyzing collaboration in food assistance networks using agent-based modeling. The University of Texas at Arlington

91. Namany S, Govindan R, Alfagih L, McKay G, Al-Ansari T (2020) Sustainable food security decision-making: an agent-based modelling approach. J Clean Prod 255:120296. https://doi.org/10.1016/J.JCLEPRO.2020.120296

92. UN-ISDR (2009) Terminology on disaster risk reduction

93. Namany S, Al-Ansari T, Govindan R (2018) Integrated techno-economic optimization for the design and operations of energy, water and food nexus systems constrained as non-cooperative games. Comput Aided Chem Eng 44:1003–1008

94. Basil M, Jamieson A (1999) Uncertainty of complex systems by Monte Carlo simulation. Meas Control 32(1):16–20. https://doi.org/10.1177/002029409903200104

95. Liu J, Li YP, Huang GH, Zhuang XW, Fu HY (2017) Assessment of uncertainty effects on crop planning and irrigation water supply using a Monte Carlo simulation based dual-interval stochastic programming method. J Clean Prod. https://doi.org/10.1016/j.jclepro.2017.02.100

96. Kadigi IL et al (2020) An economic comparison between alternative rice farming systems in Tanzania using a Monte Carlo simulation approach. Sustainability 12(16):6528. https://doi.org/10.3390/su12166528

97. Singh A (2019) Foundations of machine learning. SSRN Electron J. https://doi.org/10.2139/ssrn.3399990

98. Kumar R, Singh MP, Kumar P, Singh JP (2015) Crop selection method to maximize crop yield rate using machine learning technique. https://doi.org/10.1109/ICSTM.2015.7225403

99. Kuwata K, Shibasaki R (2015) Estimating crop yields with deep learning and remotely sensed data. https://doi.org/10.1109/IGARSS.2015.7325900

100. Bagheri M, Al-Jabery K, Wunsch D, Burken JG (2020) Examining plant uptake and translocation of emerging contaminants using machine learning: implications to food security. Sci Total Environ. https://doi.org/10.1016/j.scitotenv.2019.133999

101. Karnib A (2017) Water-energy-food nexus: a coupled simulation and optimization framework. J Geosci Environ Prot 05(04):84–98. https://doi.org/10.4236/gep.2017.54008

102. Anthony RN (1965) Planning and control: a framework for analysis. Division of Research, Harvard Business School

103. Storn R (1995) Constrained optimization. Dr Dobb's J. https://doi.org/10.1201/b18469-7

104. Namany S, Al-Ansari T, Govindan R (2019) Optimisation of the energy, water, and food nexus for food security scenarios. Comput Chem Eng 129:106513. https://doi.org/10.1016/j. compchemeng.2019.106513
105. Karan E, Asadi S, Mohtar R, Baawain M (2018) Towards the optimization of sustainable food-energy-water systems: a stochastic approach. J Clean Prod 171:662–674. https://doi. org/10.1016/j.jclepro.2017.10.051
106. Beh EHY, Zheng F, Dandy GC, Maier HR, Kapelan Z (2017) Robust optimization of water infrastructure planning under deep uncertainty using metamodels. Environ Model Softw. https://doi.org/10.1016/j.envsoft.2017.03.013
107. Kashid US (2019) Application of game theory model to selecting management strategies for optimization resources in agricultural field
108. Gal T (1980) Multiple objective decision making—methods and applications: a state-of-the art survey. Eur J Oper Res. https://doi.org/10.1016/0377-2217(80)90117-4
109. Ben-Tal A, Nemirovski A (2002) Robust optimization—methodology and applications. Math Program Ser B. https://doi.org/10.1007/s101070100286
110. Hannah LA (2015) Stochastic optimization. In: International encyclopedia of the social & behavioral sciences, 2nd edn

Printed in the United States
by Baker & Taylor Publisher Services